Hands-On AI with Java

Smart Gaming, Robotics, and More

Edwin Wise

McGraw-Hill
New York Chicago San Francisco Lisbon
London Madrid Mexico City Milan New Delhi
San Juan Seoul Singapore Sydney Toronto

The McGraw-Hill Companies

1 2 3 4 5 6 7 8 9 0 DOC/DOC 0 1 9 8 7 6 5 4

ISBN 0-07-142496-2 01-12-15

The sponsoring editor for this book was Judy Bass and the production supervisor was Pamela A. Pelton. It was set in Century Schoolbook by the Alden Group, Oxford. The art director for the cover was Anthony Landi.

Printed and bound by RR Donnelley.

This book is printed on recycled, acid-free paper containing a minimum of 50% recycled, de-inked fiber.

This book is dedicated to my wonderful wife, Marla, who lets me spend all my free time writing this stuff.

CONTENTS

ABOUT THE AUTHOR

Edwin Wise is Vice President of Development and partner at Wittlock Engineering, LLC, Austin, Texas. A software engineer with over 20 years of experience, his expertise and interests range from basic electronics and software development to microcontrollers, AI, and robots. He is the author of *Applied Robotics; Applied Robotics II*; and *Animatronics: A Guide to Animated Holiday Displays*, all published by Delmar Learning. He lives in Austin.

ACKNOWLEDGMENTS

This book could not have been written without the support of my wife, Marla. And, of course, McGraw-Hill would not have asked me to write it if I didn't have the support of you, my readers.

CHAPTER 1

Artificial Intelligence

This chapter introduces the form and format of this book and the topic of Artificial Intelligence. The bulk of this chapter sets some groundwork in the form of the display classes *JGraphPanel, JLineChart,* and *JHintonGraph*, and introduces the complex topic of validating the results of your AI projects.

Introduction

What is artificial intelligence (AI)? Even limiting our scope to the world of computer science, this is a question that has different answers to different people.

You can pick up one book on AI techniques and it will contain a hundred and one ways to perform supervised learning in backpropagation networks. A different book might list the tasks and technologies required to build expert systems using predicate logic. Other books will focus on categorization of input data using self-organizing networks. Yet another programmer may eschew all of these and use state machines to manage their problem.

On a higher level, AI topics include vision processing, speech recognition, path planning, expert problem solving systems, robot guidance systems, etc. There are few areas of computer science that are untouched by AI if you paint with a broad enough brush.

In general, AI techniques are used when a direct numerical solution is not available or feasible. AI gets to attack the "hard" problems.

The philosophical landscape of AI is littered with people debating whether machines can ever be truly intelligent. That is an interesting question, and the cognitive scientists and computer science researchers who are trying to simulate or emulate actual brains have their work cut out for them. However, if you approach AI from a practical level things become much simpler. We do not care if it is truly intelligent, only that it solves our problem. Most application developers simply want an algorithm that will solve the problem in hand, and AI techniques are good at solving hard problems.

While most books will pick one or two types of AI, this book attempts to look at most of the techniques within the AI umbrella. This means our focus is broad rather than deep. This also means that we will only be able to cover the *essence* of most techniques and not delve into all of the details. On the other hand, once you have the basics in place it becomes much easier to read and apply the dizzying variety of details available in the research papers and in other books.

The content of this book embodies my philosophy of books, the three book theory (TBT). The TBT states that, for any given technology or problem, you should have three books that address it. One book can be mistaken, confusing, or leave out important details. Three books, however, can give you a clear picture. So this book is not designed to be the *one true book*, but simply one of the three books on AI you need to get the job done. This is the overview and introduction to the building blocks of AI.

The first two chapters of this book lay some groundwork.

Chapter 3 begins exploring the outer edges of AI with we investigate reflexive control systems such as PID controls and fuzzy logic.

State-based systems provide scripted responses to stimulus. These include finite-state machines and their variations, Markov chains, and even Chatbots, explored in Chapter 4.

Many applications of AI do not look "intelligent" at all, but these problems can still be quite difficult to solve. A common problem, for example, is searching for something in a large data structure, decision tree, or problem space. The obvious methods of searching, such as visiting every possible answer to see if it is the best, can be too slow. A classic example is searching for the best next move in a chess game; there are simply too many possibilities to consider. Even though it would be *possible* to visit every single one it would not be practical. These problems are where smart search algorithms come into play, as detailed in Chapter 5.

Chapter 6 takes searching further, looking at search problems where it is not even possible to visit every possible solution. More intricate search algorithms are developed here, such as reinforcement learning and genetic algorithms.

In additional to searching techniques, "classical" AI is interested in the logical approach to thinking and problem solving, and the clever manipulation of symbols. We take a look at some of these logical techniques in Chapter 7.

Sub-symbolic techniques, however, have been the darling of the AI community for a while now. They provide an entirely different way of solving difficult categorization and prediction problems. Chapter 8 looks at neural networks and supervised neural network learning methods, while Chapter 9 expands on this base with unsupervised learning methods, the self-organizing maps.

Audience

This book is written primarily for the busy programmer though it can be used by anyone with an interest in learning how different AI techniques work.

It assumes the reader has a working knowledge of Java and is already proficient at programming. The details in this book dwell on the AI aspects of the problem and not on the way it was coded.

You should also be capable of working algebra problems, and not be frightened of mathematical symbols, such as Σ.

Purpose

This book is a field-guide to the world of artificial intelligence programming. It describes a broad range of AI tools that can each be applied to a wide array of problems.

These techniques range from those that can barely be considered intelligent, such as efficient tree searching, to classical AI symbolic techniques, to the bio-mimetic neural networks.

There are many ways to write about this kind of information, with each approach fulfilling a different need. The form of this book reflects my own deep-seated love of seeing how things work. It takes a practical approach to the subject, answering the question "how does this work?" rather than "why does this work?"

Examples

Most of this book describes the algorithms of AI, but there are also quite a few words devoted to describing the code examples. When I was learning programming I found the best teacher was existing code, to see how the principles are applied in practice. Working code can clarify a problem in a way that prose cannot. At the same time, it is important to know the theory behind the code for it to make sense. I hope I found the correct balance between theory and practice.

As the title of the book suggests, all of the code in this book is written in Java. Java is used because it is clear, simple, and portable. The downside that it is a moderately slow language is minor—once an algorithm is understood it can be adapted to the language of your choice.

The Java style used in the examples does not match the Sun Microsystems, Inc. coding conventions (as described at `java.sun.com/docs/codeconv`). I write in a mix of styles that have accreted over many years of programming.

Likewise, the documentation in the source is almost but *not quite* in JavaDoc format. I have found that Doxygen (`www.doxygen.org`) uses a more efficient commenting style and it provides marvelous documentation results. Plus, it can be applied to almost any language. I can just hear the comments now, however. "Why use a standard language like Java but then ignore the standard JavaDoc?" An excellent question, I am glad you asked. Java makes for great example code and can be run on any platform regardless of the documentation style. Doxygen is, I feel, a better format than JavaDoc—it is more readable and it does more than JavaDoc does. I am simply trying to use the best tools that I can.

All of the source and documentation can be found on the accompanying CD, as well as the website www.Simreal.com. Relevant excerpts are printed in the book, though to include all of the code in print would make the text cumbersome—there are many details of a working program that do not make for interesting reading.

Once you have the source code on your computer you can re-format it to match your own tastes using one of the many Java formatting tools, such as Jalopy (`jalopy.sourceforge.net`).

All of the code in this book was written for, and runs on, Java 1.4.1 using the JCreator Pro IDE (`www.jcreator.com`) under Microsoft Windows. Though it is not visible in the finished product, debugging support was done through the ubiquitous *System.out.println()* and, when the going got tough, JSwat (`www.bluemarsh.com/java/jswat/`).

If you want to explore the shape of the different equations in this text, I recommend the graphing tool *Equation Grapher*, found at `www.mfsoft.com/equationgrapher`.

Categories of AI

There are many different ways to divide and categorize AI systems. Some of these facets of AI are explored here.

You can separate the artificial intelligence field by application: computer vision, speech recognition, theorem proving, text parsing, adaptive controllers, and so forth. It may be more useful for a reference work to divide

AI up by technique: filters, controllers, tree search, state space search, state machines, pattern categorization, prediction in time-series data, etc.

Different AI techniques apply in different contexts. Some systems are fine when they can work offline, spending significant amounts of CPU time crunching on a problem. For real-time, online problems, different techniques may need to be used.

When an AI module has to learn or adapt to its environment, is that learning done offline, supervised by the programmer who feeds it inputs and answers, or is it online, learning as it goes? Other methods are static; they are programmed to behave in a particular manner and that is what they do, every time, all the time.

In my manufacturing applications I like the results to be repeatable and predictable, so I use deterministic methods whenever possible. When the customer runs a job and gets a weird result, I can repeat that run exactly and get the same result. But some AI techniques do not work that way; randomness enters into their calculations and the end results, while consistent across runs, may not be exactly the same. The benefit of these random (stochastic) algorithms is that they can find good solutions for very large or very difficult problems.

Whether you get exact results or not can also depend on the state space of the problem. Can each possible answer be listed and then searched (in theory if not in practice), making the problem space discrete? Or does the answer lie along a continuum, with an infinite possible number of answers, some more right than others? Sometimes you can adapt your problem to be either discrete or continuous. Sometimes you have to play the hand you are dealt.

Finally, the biggest rift in the AI system is the debate between symbolic and sub-symbolic systems. Is the problem defined using logic and inferences, symbols and manipulations of those symbols? Or is the problem solved using a network of interacting, low-level computational elements, with no explicit knowledge? The sub-symbolic (also known as connectionist) solutions are typically based on biological systems.

As you read through the various algorithms in this book, notice which categories they fit:

- *Task:* Filtering, control, searching, classification, prediction, or optimization?
- *Style:* Symbolic, connectionist, or collaborative?
- *Problem type:* Discrete or continuous?
- *Solution type:* Deterministic or stochastic?

- *Learning:* Static or adaptive? If adaptive, online or offline learning?
- *Execution:* Fast or slow? Offline or online?

This list is not comprehensive. For example, it is missing the architecture dimension; is the solution a sense–plan–act cycle, or is it a set of experts working in parallel, reacting to their environment in a behavioral system? The list goes on.

Validation

For most deterministic problems, validating your solution is a simple enough task. You present the program with a set of data and check the results against known answers. For example, searching a data structure is known to work when it finds an answer when an answer exists, and it does not find an answer when one does not exist.

But even then, you may want to quantify the results of the search in terms of how *quickly* the answer was found.

Once you move into the realms of stochastic solutions (those that involve randomness), you will get degrees of "rightness" or some percentage of right answers. In these cases, you need to apply statistical tools to get an idea of whether your solution is *right enough* for production work, or if you need to develop it further.

Finally, when you are adjusting and improving your AI code, it helps to know whether the change you just made has improved or degraded your algorithm, or if the change was not significant either way.

While this is not a book on statistics, and cannot go into much depth on that subject, we touch on some statistical techniques you can use as you explore different AI solutions.

In the process of exploring various validation techniques, we lay down some basic Java code for displaying information. Visual feedback is an excellent way to get a gut feeling for how something is behaving, and these classes come in handy later as we actually do work.

Statistics

There are two kinds of statistical techniques, descriptive and inferential.

Descriptive statistics provide numbers that can be used to characterize a set of data. In this case, we have all of the information in hand (the entire population) and are simply trying to find a concise description for it. One such number is the mean μ, which is the average value for the population as defined by:

$$\mu = \frac{\sum_{i}^{N} X_i}{N} \qquad \textbf{1-1}$$

where X are the numbers being summed and N counts the number of X's.

Another such value is the standard deviation σ that describes how focused the values X are around the mean:

$$\sigma = \sqrt{\frac{\sum_{i}^{N} (X_i - \mu)^2}{N}} \qquad \textbf{1-2}$$

And so on and so forth. We will be describing and using a number of statistical values like this later, but will not spend too much time explaining the significance of them all. If you are curious, any good statistics text can fill in the details.

Of course, you can almost never use the data for the entire population, such as the result of every possible run of an algorithm or the value of every possible optimization on data. Inferential statistical techniques use additional hand-waving to extrapolate information from a small sample of the total population into information about the population as a whole.

If you are interested in a better understanding of statistics, as well as information on other techniques not mentioned below, find a book or two on the subject. I can recommend Spatz (1997) for a start, and you can move on from there.

Validating Filtering and Predictive Systems

Two classes of problem share a common diagnostic. Both filters and predictors can be quantified with the signal-to-noise ratio. This comes not from the world of statistics but from signal processing.

Filters The purpose of a filter, for this example, is to take a signal with added noise and process it so that there is more signal and less noise.

The signal might look like:

$$t_i = s_i + n_i \tag{1-3}$$

where t_i is the total signal, s_i is the pure signal, and n_i is additive noise, all for signal time index i. Validation tests would use both a known signal and a known noise source, though in practice the filter would not know about either of these in advance. The purpose of validation is to see how the system will work in theory and, hopefully, it will continue to work that way in practice.

The total signal power S for all signal entries across time T is:

$$S = \frac{\sum_{i}^{T-1} s_i^2}{T} \tag{1-4}$$

and likewise the total noise power N is:

$$N = \frac{\sum_{i}^{T-1} n_i^2}{T} \tag{1-5}$$

Finally, as you would expect, the signal-to-noise ratio is *S/N*, but that is only part of the story.

The ratio of two powers is typically measured in the decibel, which is defined as 10 $\log_{10}$ of the ratio in question:

$$S/N = 10 \log_{10}\left(\frac{S}{N}\right) \tag{1-6}$$

Each doubling of the signal-to-noise ratio is a ten-times increase in signal over the noise. So a *S/N* of 10 means that the signal is 10 times as powerful as the noise, and a *S/N* of 20 means the signal is 100 times as powerful as the noise.

Using the *S/N* values you can compare the effects of different filters.

Predictors A prediction system is trying to determine the next value of a signal given the current and past values of that signal. To determine the *S/N* for a predictor, the noise is the error of the predicted values relative to the actual signal value.

While the signal power S is still calculated for the true signal (Equation 1–4), the noise is calculated from the error in prediction:

$$N = \frac{\sum_{i}^{T-1} (s_i - p_i)^2}{T} \qquad \textbf{1-7}$$

where p_i is the prediction for a given time slot.

Now, the *S/N* formula can be compacted a bit, since the $1/T$ factors cancel out, giving:

$$S/N = 10 \log_{10}\left(\frac{\sum_{i}^{T-1} s_i^2}{\sum_{i}^{T-1} (s_i - p_i)^2} \right) \qquad \textbf{1-8}$$

Code: Strip Chart Display

A scrolling strip chart would be perfect for visualizing the data in Equation 1-3, or any other time-varying information.

This strip-chart program introduces three classes and uses the Java standard classes *JPanel* and *JFrame*, as shown in Figure 1-1. *Signal* is the test program itself, and it generates a signal, noise, and signal+noise. *JLineChart* is the line chart and it, in turn, is a member of *JGraphPanel*. *JGraphPanel* provides some basic services. In addition to this class hierarchy, the utility class *MathConst* is introduced. This test code is described here, one class at a time.

These classes, and many others in the *aip* libraries, are grouped together under the *AIP* project as defined by *AIP.jcp*. Even if you do not use JCreator, the *jcp* file is easy to read and understand.

MathConst.java The *MathConst* class provides tools for manipulating floating point numbers. All of the projects that I work on use this class or one much like it. Most of its methods provide tolerance-safe comparisons for numbers, since when you work with floating point values direct comparisons are not to be trusted. This is due to floating point drift during

Figure 1-1 Signal and JLineChart class diagram

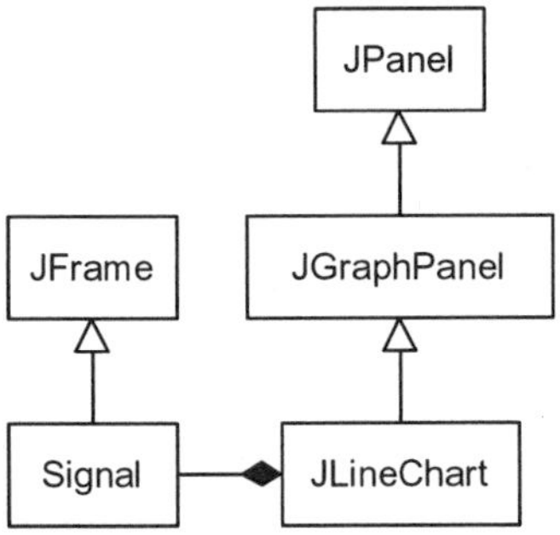

calculations, where the least significant digits wander around. The default tolerance is SMALL, defined as 0.000001, but this can be overridden as needed.

The code itself is self-explanatory, and includes methods to test whether numbers are zero, undefined, positive, negative, and so forth. There are also tests to compare two numbers, and to convert between degrees and radians. Finally, you will find a number of handy numeric constants at the end.

JGraphPanel.java

aip.display.JGraphPanel *JGraphPanel* is not usable by itself, but requires a sub-class to fill in its various data tracking and display tasks. The one thing *JGraphPanel* does is manage the graph's labels and their colors, since these things are common across all of the display classes.

Each *JGraphPanel* display is composed of layers of information. The top layer consists of any data and axes labels, and the data axes and tick marks themselves. This top layer is repainted by *JGraphPanel* and its children from scratch every time *paintComponent()* is called.

The lower layers are data layers. There may be more than one of these. The children of *JGraphPanel* manage any data to be displayed. These layers may be re-drawn with each paint, or they may only be a reflection of past data. *JGraphPanel* keeps an *Image* object handy to store any persistent visual information. For example, the *Image* object *m_data_image* is used by *JLineChart* to minimize the amount of drawing that needs to be done each paint.

A summary of *JGraphPanel* is given in Table 1-1.

Looking at the code on the disk, the first thing that may leap to attention is the Doxygen commenting style. There are two conventions used here that bear explanation. The first are the comments that control grouping:

Table 1-1

JGraphPanel summary

Construction

```
                JGraphPanel(int dx, int dy)
```

Data Feed

```
         int addLayer (String label, Color color)
        void setRange (int dim, double min, double max)
         int getDimension ()
        void addDatum (int layer, double y)
        void addDatum (int layer, double x, double y)
        void addDatum(int layer, double x, double y, double a)
abstract void done ()
```

Get/Set

```
         int getLayerCount ()
      String getLabel (int layer)
        void setLabel (int layer, String label)
       Color getColor (int layer)
        void setColor (int layer, Color color)
         int getStyle ()
        void setStyle (int style)
         int getWidth ()
         int getHeight ()
       Image getDataImage ()
```

Painting

```
        void paintComponent (Graphics g)
        void paintLabels (Graphics g)
```

```
/** @name Data Feed
 *  Provides stubs for the various forms of data that we may
    provide. The
 *  default addDatum implementations simply throws an exception,
    so
 *  children only have to override the ones they care about.
 */
//@{
  ... methods
//@}
```

The *@name* and following comment describe the grouping. The *//@{* and *//@}* provide the boundaries for the grouping itself. This provides a nice functional organization for the method documentation.

The second is the way parameters are documented:

```
public void addDatum(
   int layer,        ///< Layer this datum is associated with.
   double x,         ///< X-axis component
   double y,         ///< Y-axis component
   double a )        ///< Attribute of this datum
   throws IllegalAccessException
{
   ... code
}
```

The Doxygen style allows us to comment on parameters on the parameter itself, which avoids the redundancy and possible synchronization errors of standard JavaDocs.

The working methods in *JGraphPanel* are in the Data Feed block. The *addDatum()* methods are all stubs to be replaced by sub-classes. The *addLayer()* method actually does work; it records a layer name and color so the children of this class do not have to.

Selected methods are described here.

JGraphPanel(int dx, int dy) There is only one constructor for *JGraphPanel*, and it specifies the width d*x* and the height d*y* of the graph's content area, in pixels. All data added to the graph is scaled to fit in this display area.

addLayer(String label, Color color) *addLayer()* is used to create a new display overlay of the graph. At least one layer must be created for the graph to have any purpose.

setRange(int dim, double min, double max) While the graph is self-scaling, you can also pre-set the minimum and maximum values you expect it to receive. The *dim* parameter defines which dimension you are setting the extents of, starting with zero. For *JLineChart* there is only one dimension, zero. Other graphs may have two or more dimensions.

addDatum(int layer, ...) There is a unique *addDatum()* method for each graph dimensionality; one-dimensional, two-dimensional, and attributed two-dimensional. For each of these, the *layer* parameter defines which overlay layer we are adding the datum to. How this datum is then displayed depends on the specific child class.

done() For many types of graphs, if there is only one layer then that layer will be re-drawn as soon as you call *addDatum()*. This is for convenience only, and can be removed if it offends you. For multi-layer graphs, you must call *done()* explicitly before the layers re-draw.

paintComponent(Graphics g)
paintLabels(Graphics g) *paintComponent()* is called by the environment when it is time to re-draw this component. *paintComponent()* calls *paintLabels()* in turn, to draw the graph labels.

JLineChart.java

aip.display.JLineChart *JLineChart* is a form of *JGraphPanel* that emulates a strip recorder, where a scrolling piece of paper has marks placed

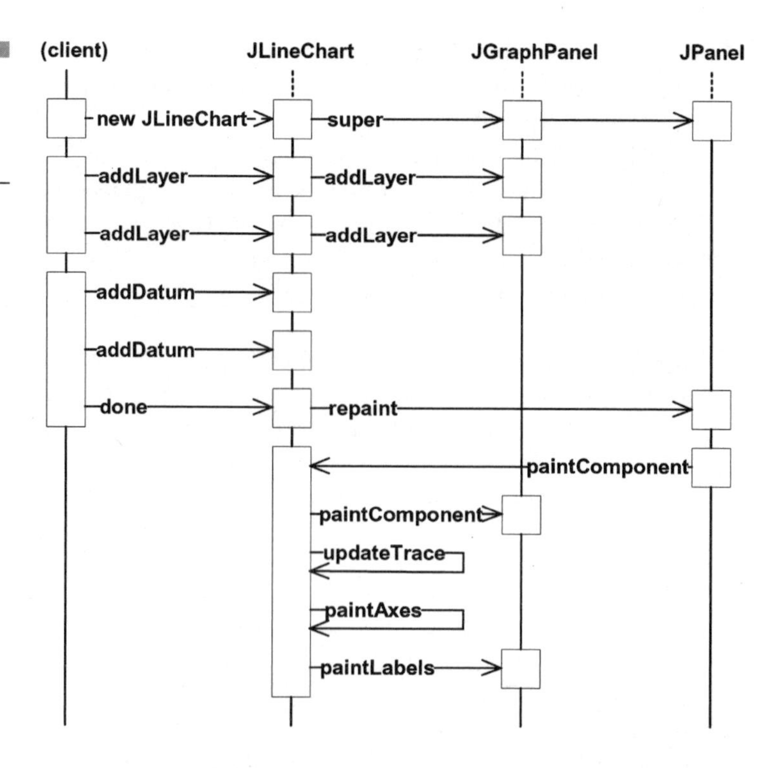

Figure 1-2 JLineChart interaction diagram

on it by pens floating on its right-most edge. Think of it as a virtual EEG peering into the brain of your program.

This chart only draws the most current data point, at the right edge of the chart, unless there has been a re-scaling event. While the chart's client can set the starting range for the graph, it is also self-scaling and adjusts to data as it comes in.

The basic behavior of *JLineChart* is shown in Figure 1-2. There are four phases of operation.

First, of course, a *JLineChart* object must be created.

Then one or more data layers are added to the chart. When a new data point, or set of data points, is available they are added to the chart.

If there is more than one layer, *done()* must be called to signal the chart to update itself.

Finally, on *done()* or any other re-paint, the base *JPanel* component asks us to paint ourselves, whereupon the *JLineChart* and *JGraphPanel* interact to create a valid image of the data.

A summary of *JLineChart* is given in Table 1-2.

Most of the methods in *JLineChart* mirror their counterparts in *JGraphPanel*. Those that are unique to *JLineChart* are also not of much interest to outside code.

Table 1-2

JLineChart summary

```
Construction
     JLineChart (int dx, int dy)

Data Feed
 int addLayer (String label, Color color)
void setRange (int dim, double min, double max)
 int getDimension ()
void addDatum (int layer, double y)
void done ()
void rescale ()
 int toPixel (double val)

Painting
void paintComponent (Graphics g)
void paintAxes (Graphics g)
void paintTrace (Graphics g, int layer)
void updateTrace (Graphics g, int layer)
```

JLineChart(int dx, int dy) The *JLineChart()* constructor mirrors its parent's constructor.

addDatum(int layer, double y) The *X* position of each datum in this strip chart is fixed at the right-hand edge of the graph. When a new datum is added, a line of the appropriate color is drawn from the previous *Y* position to the newly specified *Y*. The net effect is that of a continuous strip chart.

Signal.java

aip.app.signal.Signal Finally, we can approach the *Signal* application itself. This project is defined in the *Signal.jcp* project.

Signal is an extension of *JFrame* and executes starting with the *main()* static method. This bootstraps the operation by allocating a *Signal* object and executing its *assemble()* and *run()* methods. This startup sequence is generic and immutable across the various applications from this book.

The first method called, *assemble(),* simply creates the various objects and display layers.

The work happens in *run()*, where there are three phases of operation. First, three signals are calculated for a given point in time. Second, they are sent to the *JLineChart*. And finally, it sleeps for 25 milliseconds.

This final sleep is important for three reasons. One is that it allows our application to "play nice" and not hog all of the CPU. The second is that it evens out the timing of the application, so the animation is nice and smooth. The third and possibly most important reason is, in giving the operating system a chance to process events it allows *repaint()* and therefore *paintComponent()* a chance to operate.

Table 1-3

Signal summary

Public Methods	
	`Signal ()`
`void`	`assemble ()`
`void`	`run ()`
Static Public Methods	
`void`	`main (String args[])`

A summary of *Signal* is given in Table 1-3.

Signal() The *Signal()* constructor simply adds a window listener, so we know when the application is being closed by the user.

assemble() About half of *assemble()* is dedicated to putting together the user interface of this test application. The rest of it creates a *JLineChart*, creates the layers in the chart, and adds it to the UI.

```
content = getContentPane();
//
//
m_chart = new JLineChart(300, 100);
Border border = BorderFactory.createBevelBorder BevelBorder.LOWERED);
m_chart.setBorder(border);

m_chart.addLayer("One", Color.red);
m_chart.addLayer("Two", Color.blue);
m_chart.addLayer("Three", Color.darkGray);

content.add(m_chart);
```

run() This is the main loop of the application. It generates some signals, adds them to the chart, and then sleeps briefly so things run smoothly.

```
while(true)
{
   double v1 = Math.sin(t*10);
   double v2 = Math.random() * NOISE*2.0 - NOISE;
   double v3 = v1 + v2;

   m_chart.addDatum(0, v1);
   m_chart.addDatum(1, v2);
   m_chart.addDatum(2, v3);
   m_chart.done();

   t += dt;
   // Play nice
   try {
      Thread.sleep(25);
   } catch(Exception e) {}
}
```

Execution ends when the user dismisses the display window.

Validating Optimizations

There are two closely related kinds of optimization we address with statistical analysis.

On one hand, there are algorithms that work to optimize some value like path distance in the traveling salesman problem, or material use in a pattern-nesting system. For these optimization algorithms there will be some value that indicates the score for a given result. Or the score could be the number of steps, or the time taken, to perform the operation. Changes in the algorithm may result in changes in that score, and we need to be able to tell if that change is significant or not.

On the other hand, you may have an algorithm that has one or more parameters that control and tune its operation. As you tune and adjust these parameters you need to be able to test if the system is getting better or not. Just as in the previous example, there will be some related score for the operation.

This score is the important piece. For statistics to do us any good, the score needs to be a value that changes with each run. In a deterministic system, the score may be exactly the same for a given set of data, but it would be different for different sets of data, perhaps in a non-intuitive way. For a stochastic system the score will be different on every run, regardless of whether the data it is running against has changed or not.

The first thought is that it should be obvious whether a change in the optimization has improved it or not. Just look at the scores!

Unfortunately, things are rarely that simple. Different runs will give different scores for one reason or another. And a sample from the population of all possible runs will give a range of different scores. Within the many constraints that statistical methods impose, this randomly selected sample from the greater population will fall into a "normal" distribution, which is described by that most famous of curves, the bell curve (Figure 1-3).

The peak of the bell curve is at the mean. The two tails, while trailing off to infinity, can be considered to end at ±3 SD from the mean, though tables will list values out to ±4. Ninety-five percent of the samples are within two deviations.

The normal distribution can be thought of as a series of bins that you store results in. The left-most bin has, for example, the smallest results (or range of results if they are not integral), and the right-most bin has the largest results or range of results. When you run a test, you check that specific result to see which bin it falls into and increase the level in that bin. If you run often enough, and the results are really a random

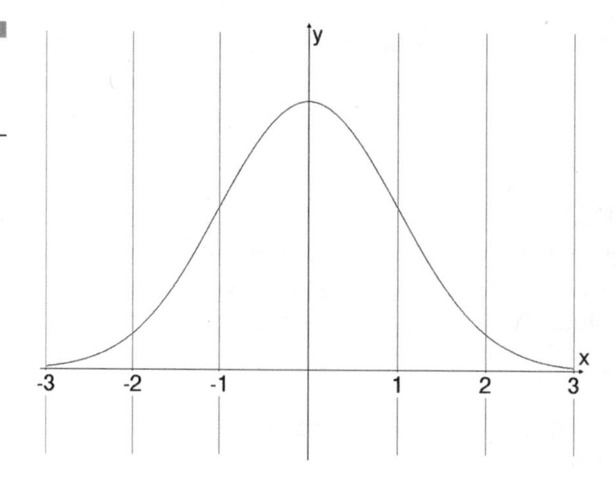

Figure 1-3 Normal distribution curve

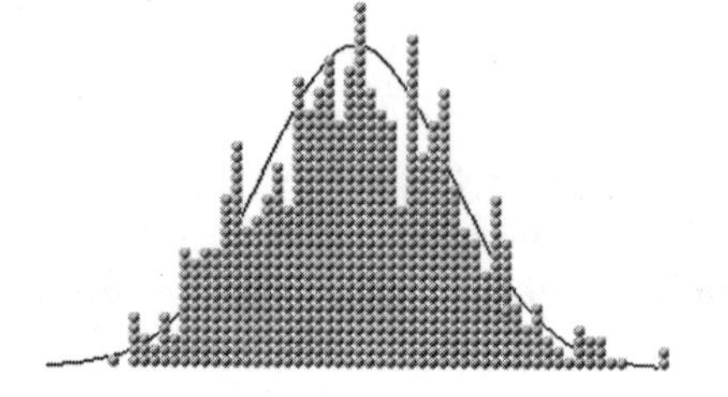

Figure 1-4 Bin-filling in the normal curve

sampling from the population of all possible results for this problem, the bins will fill up along the lines of the curve as shown in Figure 1-4.

As you can see, the results are not perfect, but they do indeed follow the curve of the normal distribution.

Now assume that you made some changes to the algorithm and run the tests again. The new run (black) is shown together with the original run (grey) in Figure 1-5. What does it mean? It looks better, but is this improvement within the random variation expected by the algorithm, or is it truly a better version?

There are two problems to solve. The first is how to take a set of raw data, the results of your test runs, and turn them into a concise statistical description. The second is how to compare the two sets of run results to see if they are significantly different or not.

Raw Data into Statistical Information The two values used to describe a distribution are its mean and its standard deviation. The formulas to calculate these values for a population were given in Equations 1-1 and 1-2. When you calculate the mean and standard

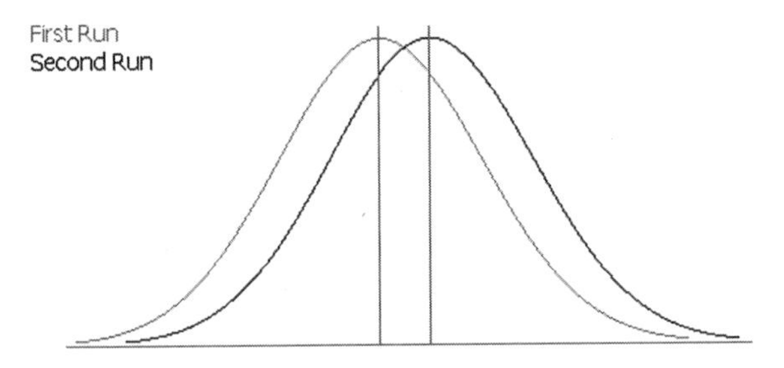

Figure 1-5
A second run with different results

deviation for a sample set, the equations are almost but not quite exactly the same as shown those in Equations 1-1 and 1-2.

$$\bar{X} = \frac{\sum_{i}^{N} X_i}{N} \quad \textbf{1-9}$$

$$S = \sqrt{\frac{\sum_{i}^{N} (X_i - \bar{X})^2}{N}} \quad \textbf{1-10}$$

The mean of a sample is now $\bar{X}$, pronounced "ex-bar", and the standard deviation for a sample is S. Since we are now working with samples, however, we have one more measure, the estimate s of the standard deviation of the population is:

$$s = \sqrt{\frac{\sum_{i}^{N} (X_i - \bar{X})^2}{N-1}} \quad \textbf{1-11}$$

The details of *why* we divide by $N - 1$ instead of N can be found in better statistical texts at a bookstore or library near you. Note, however, that s will be a larger value than S, due to the smaller divisor, making it a somewhat less sure value. This is appropriate, since while the sample reflects the population, it does not represent it exactly.

The more sample scores X that you use, the better the sample will represent the population. How many samples you use can vary from 2 to an infinite value. When you have a choice, 30 samples is a good number.

Notice again that the more samples you take the more precise your estimates are, and the s statistic works best when there are a huge number of samples. There is another statistic that explicitly takes the number of samples into account, giving broader results when presented with fewer samples. This is the t statistic, meaning, I assume, the statistic after s.

Later we use a variation of the t statistic to determine if the two samples are from the same population or not.

The t statistic, fittingly enough, falls into the curves known as the t-distribution. There is a unique t-distribution for every possible sample size greater than two. Each distribution is known by its degree of freedom (df). For our purposes, the value of df can be that defined in Equation 1-12.

$$\mathrm{df} = N - 1 \qquad \textbf{1-12}$$

Figure 1-6 shows some representative t-distributions, approaching the normal curve as df approaches infinity. The actual formula for the t-distribution is very complicated, and unless you have specialized statistical software your best bet is to look up the relevant values in a table, such as Table 1-4 at the end of this section.

To use the t-distribution, you need to calculate the t score for your sample:

$$t = \frac{\bar{X} - s}{s_{\bar{X}}} \qquad \textbf{1-13}$$

This uses a new value $s_{\bar{X}}$ which is the standard error of the sample, and is defined as:

$$s_{\bar{X}} = \frac{s}{\sqrt{N}} \qquad \textbf{1-14}$$

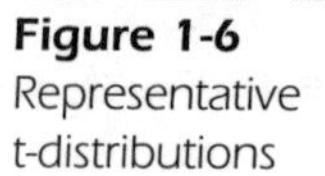

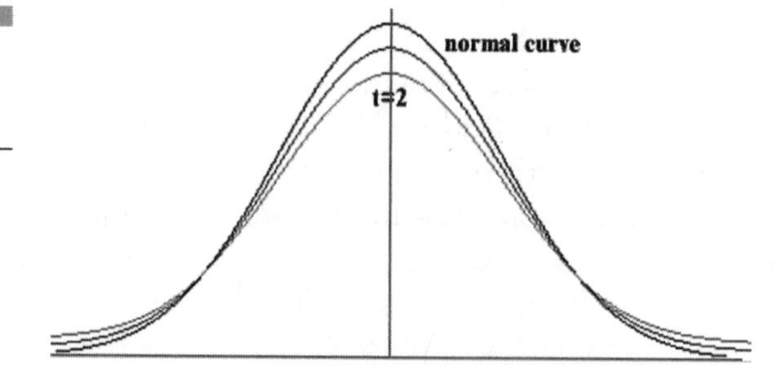

Figure 1-6 Representative t-distributions

Table 1-4

t-distribution at 95% confidence

df	*t* at 0.05 significance
1	6.314
2	2.920
3	2.353
4	2.132
5	2.015
6	1.943
7	1.895
8	1.860
9	1.833
10	1.812
11	1.796
12	1.782
13	1.771
14	1.761
15	1.753
16	1.746
17	1.740
18	1.734
19	1.729
20	1.725
21	1.721
22	1.717
23	1.714
24	1.711
25	1.708
26	1.706
27	1.703
28	1.701
29	1.699
30	1.697
40	1.684
60	1.671
120	1.658
∞	1.645

Equations 1-12 through 1-14 provide a context for the next section, comparing distributions.

Comparing Two Distributions The first step, of course, is to collect the data from both versions of the solution. The unmodified solution can be solution A and the version we have changed and are testing can be solution B.

Technically what we are going to do is a two-sample hypothesis test. The first hypothesis, H_0, is known as the null hypothesis and states that the two samples are essentially the same. The second hypothesis, H_1, is the alternate hypothesis and it states that the two samples are different within some level of significance.

The traditional level of significance is $\alpha = 0.05$, which is a 95% confidence interval. This leaves a 5% chance that you got the wrong answer, which is not too bad.

Equation 1-13 calculates a t score for a single distribution, which can be used for one-sample hypothesis tests, which are not explored here. This would be used to test if your sample has a particular mean or not. In a two-sample test, we are trying to discover if the two samples come from the same population. Do they represent a population that has the same mean? Or do they come from two different populations? This requires a different calculation for the t score:

$$t = \frac{\bar{X}_A - \bar{X}_B}{s_{\bar{X}_A - \bar{X}_B}} \qquad \textbf{1-15}$$

This in turn requires the new value $s_{\bar{X}_A - \bar{X}_B}$ which is the standard error of a difference. With the simplifying assumption that $N_A = N_B$, this can be calculated as:

$$s_{\bar{X}_A - \bar{X}_B} = \sqrt{s_{\bar{X}_A}{}^2 + s_{\bar{X}_B}{}^2} \qquad \textbf{1-16}$$

And in the two-sample case, there is also a different calculation for degrees of freedom:

$$\text{df} = N_A + N_B - 2 \qquad \textbf{1-17}$$

Now go down the rows of Table 1-4 until you find your df value and you will see which t value is considered significant. If the t score for your two samples is greater or equal to the number in the table, then you can reject H_0 with 95% certainty, meaning the samples are very likely from different populations.

Validating Classification

Many AI and especially neural network systems classify input data. Classification can be simple yes/no tests, or can assign data into any number of categories. In fact, the tests done above comparing two distributions can be considered a classification problem, dividing the test pair into the categories "same" and "different".

There are several tools you can use to help determine how well your categorization system is working, some of which are explored here. We divide our discussion into two parts, with the first part looking at two-category systems and the second part looking at systems with more than two categories.

Two-Category Systems For any test where you decide if something is true, or part of a category or not, there are two kinds of errors you can make. It goes without saying that there are names for these errors.

- A Type I error is where something belongs to a category but you reject it anyway. The probability of making Type I errors often represented by α.
- A Type II error is where something does *not* belong to a category but you include it anyway. The probability of making a Type II error is often represented by β.

There are actually four possible results of a categorization, also called a prediction, as shown in Table 1-5. While there are two categories, the test is for whether a sample is in category one or not. Category two is made up of everything that fails the test.

- a is the number of times it *correctly* predicts category *one*,
- b is the number of times it *incorrectly* predicts category *two*,
- c is the number of times it *incorrectly* predicts category *one*.
- d is the number of times it *correctly* predicts category *two*.

With regards to category one, b is a Type I error and c is a Type II error.

There are a number of standard terms defined from the entries in this matrix.

Table 1-5

2-category matrix

		Predicted	
		True	**False**
Actual	True	a	b
	False	c	d

- Accuracy (AC) is the proportion of correct predictions:

$$\mathrm{AC} = \frac{a+d}{a+b+c+d} \tag{1-18}$$

- True positive rate (TP) or recall rate, is the proportion of positive cases that were predicted correctly:

$$\mathrm{TP} = \frac{a}{a+b} \tag{1-19}$$

- False positive rate (FP) is the proportion of negative cases that were incorrectly predicted as positive:

$$\mathrm{FP} = \frac{c}{c+d} \tag{1-20}$$

- True negative rate (TN) is the proportion of negative cases that were predicted correctly:

$$\mathrm{TN} = \frac{d}{c+d} \tag{1-21}$$

- False negative rate (FN) is the proportion of positive cases that were incorrectly predicted as positive:

$$\mathrm{FN} = \frac{b}{a+b} \tag{1-22}$$

- Precision (P) is a different proportion of positive cases that were properly predicted:

$$P = \frac{a}{a+c} \tag{1-23}$$

These ratios work best when there are a similar number of positive and negative cases in the test.

The different ratios can be used to measure the quality of your predictions. They can be especially useful if the cost of different predictions, especially prediction errors, is different.

A classifier will often have a threshold value, α, for acceptance into a category. When α is zero, no sample will be considered a member of the category regardless of its actual state or the results of the test. In this case, the TP rate is zero and the FP rate is also zero, since there are no positives at all. When α is one, all samples are considered positive, with a TP and FP rate of one. Moving α between zero and one will provide different ratios of predictions.

When you graph the TP rate against the FP rate while varying α from 0.0 to 1.0, you get a picture of the classifier and how it behaves at different α thresholds. This graph is the receiver operating characteristics or relative operating characteristics (ROC) curve, as shown in Figure 1-7.

The diagonal line from [0,0] to [1,1] represents random guessing. An ROC curve above this line is performing better than chance, and a curve below this line is worse than chance. Of course, if your curve is operating worse than chance you can simply reverse the meaning of its decisions to flip the curve over to the other side.

A perfect classifier operates at [0,1]. Odds are good that your classifier is operating somewhere between random chance and perfection.

The ROC curve gives you a "feel" for how your classifier is working, which is nice. But the real value comes in using it to determine which α level would be the best one to operate with.

The best α level on the curve depends on the cost of a false positive versus the cost of a false negative, as well as the proportion of positive cases to negative cases. For any given point [FP, TP] on the ROC graph, the total cost C of that point is:

$$p = \frac{a+b}{a+b+c+d} \quad \textbf{1-24}$$

$$C = (1-p)\mathrm{A} * FP + p\mathrm{B} * FN \quad \textbf{1-25}$$

where A is cost of one false positive and B is the cost of one false negative.

Figure 1-7
ROC curve

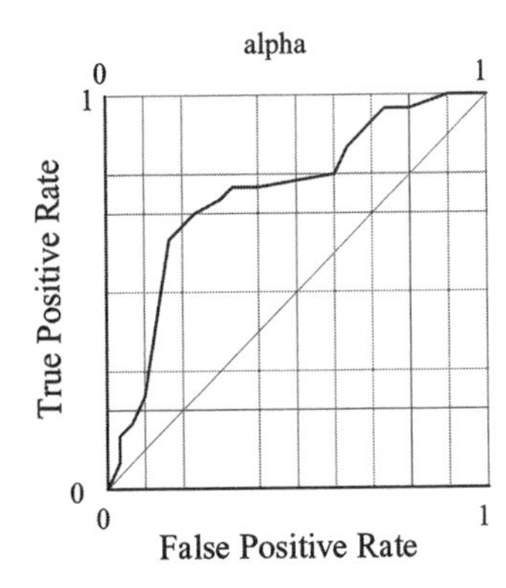

For any given set of parameters, there will be a set of straight lines of equal costs C in the graph, the isocost lines. When there are an equal number of positive and negative cases, and the cost for a false positive is the same as a false negative, these will be at a 45° angle. The best operating point is then the point on the curve that is both tangent to an isocost line and is closest to the [0,1] point, as shown in Figure 1-8.

Figure 1-9 shows how this changes when the overall cost of a false positive is three times that of a false negative.

A different way to account for the cost of a point is to take its Euclidean distance from the [0,1], the point where all classifications are perfect. The goal in this case is to minimize C:

$$C = 1 - \sqrt{FP^2 + (1 - TP)^2} \quad \textbf{1-26}$$

where [FP,TP] is a point on the ROC curve.

If you are comparing two classifiers, you can compare the areas under each of their curves. This only gives a "rule of thumb" on their relative

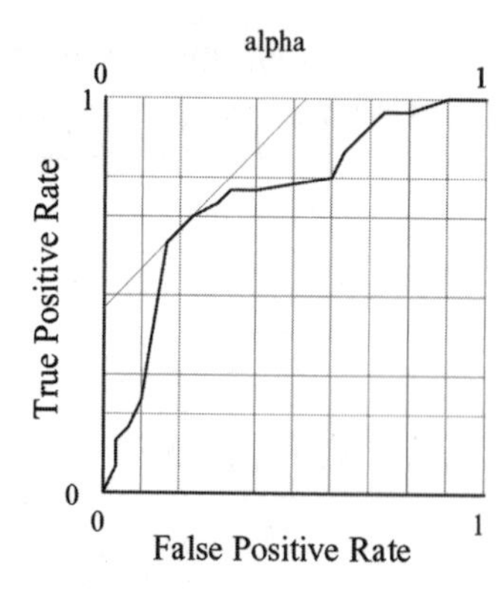

Figure 1-8
Equal costs

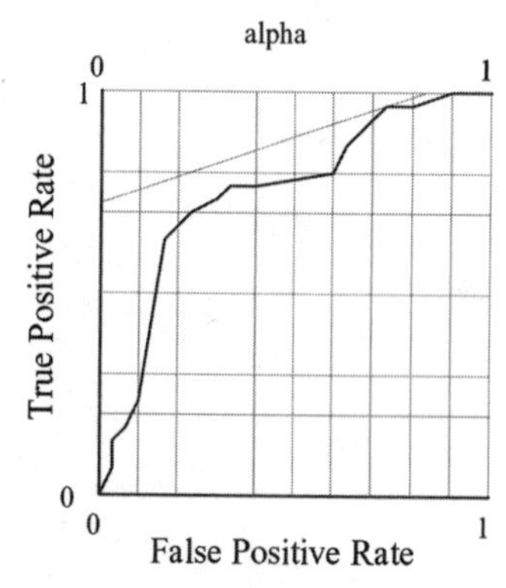

Figure 1-9
FP costs 3x FN

merits, since the shape of the curve (and the operating point you use on the curve during normal operation) makes a big difference.

More than Two Categories Though you could analyze an N-category classifier under the previous schemes by reducing the range of its results to Correct and Incorrect, it is also possible to get a direct picture of the classifier's actions.

This discussion could get messy in the abstract, so let us invent an example in automatic speech recognition. Say you had a system that could classify a sound from speech as a stop, fricative, sonorant, or silence. For a sound in each of these four categories, the classifier can identify it correctly or falsely assign it to one of the other categories. So to begin, we create a grid that lists the number of times each condition occurs during testing (Table 1-6).

You can see a definite trend of large numbers along the diagonal, which seems to indicate that this classifier is doing a decent job. However, the numbers in each row are all in their own scaling system. There were twice as many sonorant sounds classified than there were stop sounds, so a value of two in the sonorant row means something different than a two in the stop row. So the first step is to normalize the matrix by dividing each count by the sum of its row, as shown in Table 1-7.

Now each cell has a tidy value from 0.000 to 1.000 that indicates how often that categorization occurred.

Another way to normalize the diagram is to scale each entry by the total number of samples taken, so that each cell indicates the chance of that condition occurring across the test (Table 1-8). The cell values can then be multiplied by the cost of that specific cell occurring, which may be zero for correct classifications and some large number for heinous mistakes. These costs may be summed across the table to get the expected cost of running this classifier on a sample. The goal, of course,

Table 1-6

Confusion matrix for speech sounds

		Predicted			
		Stop	**Fricative**	**Sonorant**	**Silence**
Actual	Stop	135	10	2	3
	Fricative	27	189	4	5
	Sonorant	3	2	285	10
	Silence	3	4	1	83

Table 1-7
Confusion matrix normalized by row

		Predicted				
		Stop	Fricative	Sonorant	Silence	Row
Actual	Stop	135, 0.900	10, 0.067	2, 0.013	3, 0.020	150
	Fricative	27, 0.120	189, 0.840	4, 0.018	5, 0.022	225
	Sonorant	3, 0.010	2, 0.007	285, 0.950	10, 0.033	300
	Silence	3, 0.033	4, 0.044	1, 0.011	83, 0.912	91

Table 1-8
Confusion matrix normalized by table

		Predicted			
		Stop	Fricative	Sonorant	Silence
Actual	Stop	0.176	0.013	0.003	0.004
	Fricative	0.035	0.247	0.005	0.007
	Sonorant	0.004	0.003	0.372	0.013
	Silence	0.004	0.005	0.001	0.109
		766 samples			

is to find a classifier with the minimum cost. Note that the cost factors for the various mistakes heavily affect the ultimate outcome.

These normalized forms can be used to compare different classifiers, or you could draw the results in a Hinton Diagram (Figure 1-10) for a nice visual representation.

Code: Hinton Diagram Display

A Hinton Diagram is a grid of dots, where each dot has a size and color based on the value at that position in the grid. The size goes from zero for a zero value, to a dot that fills its grid space for a value magnitude of 1.0. The color is based on the sign of the value. For example, the data in Tables 1-7 and 1-8 are illustrated by the Hinton Diagram in Figure 1-10.

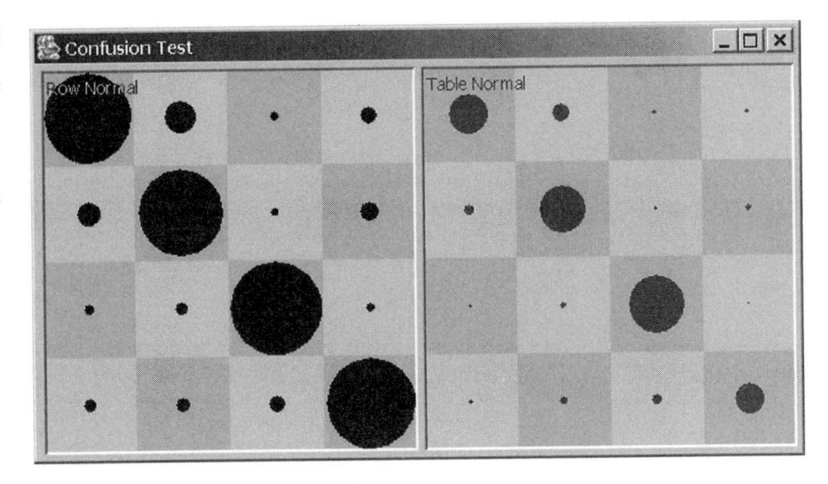

Figure 1-10
Hinton diagram of normalized confusion matrices

JHintonGraph is a class that displays Hinton diagrams. Like *JLineChart*, *JHintonGraph* is based on the class *JGraphPanel* as shown in Figure 1-11. The behavior is almost the same as shown in Figure 1-2, though there are a few oddities explained below.

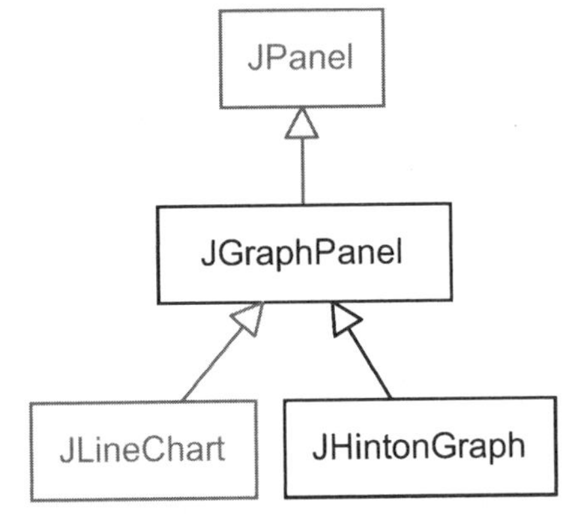

Figure 1-11
JHintonChart class diagram

JHintonGraph.java

aip.display.JHintonGraph *JHintonGraph* is different from *JLineChart* in that it is definitely not self-scaling. Indices into the graph must be within a pre-set range, which means that the client code must call *setRange()* for both axes before data is added. Because of its strongly gridded nature, the *X* and *Y* values should all be nice integral values.

Table 1-9

JHintonGraph summary

```
Construction
              JHintonGraph (int dx, int dy)
        void  create ()

Data Feed
         int  addLayer (String label, Color color)
        void  setRange (int dim, double min, double max)
         int  getDimension ()
        void  addDatum (int layer, double x, double y, double a)
        void  done ()

Painting
        void  paintComponent (Graphics g)
        void  paintData (Graphics g)

Public Attributes
final int     StyleDot = 0
final int     StyleSquare = 1
```

The graph may have one or two layers. The first layer holds the actual data, while the optional second layer provides an alternate color for negative values. If a second color is not specified, *Color.red* is used.

The bulk of the effort is done in *paintData()*. Iterating through the grid of data, a cell is re-drawn only if its value has changed. In that case, some light Boolean slight of hand is done to determine the background shade for that cell and then a dot of the appropriate color, size, and style is painted over it.

A summary of *JHintonGraph* is given in Table 1-9.

JHintonGraph(int dx, int dy) The constructor is not very interesting. The real construction work is actually done in *create()*.

create() When the very first call to *addDatum()* is made, *create()* is called. This method determines the size of each "cell" in the graph and allocates the grids that store the datum values.

Note that the current data is in one two-dimensional array, and the previous values of the grid are in another array. This allows the display code to determine which values have changed to minimize drawing.

setRange(int dim, double min, double max) The *JHintonGraph* must know how many rows and columns to allocate, so before any data is added the range of both the *X* (*dim* = *0*) and *Y* (*dim* = *1*) axes must be specified.

addDatum(int layer, double x, double y, double a) The *X* and *Y* parameters determine the column and row of the cell we are setting. Though defined as *double* values because of the *JGraphPanel* parent class, these are interpreted as zero-based indexes.

The *a* parameter is the attribute of the cell to set. A positive value displays in the primary layer's color and a negative attribute value displays in red or the secondary layer's color. The attribute value is expected to fall in the range [–1.0 .. 1.0].

paintData(Graphics g) Normally, this method would not require any special attention. However, there is a subtle trick being performed here.

```
val = Math.sqrt(Math.abs(val));
int dx = (int)(m_cell_dx*val);
int dy = (int)(m_cell_dy*val);

switch (getStyle())
{
case STYLE_SQUARE:
   g.fillRect(ctr_x-(dx/2), ctr_y-(dy/2), dx, dy);
   break;
case STYLE_DOT:
   g.fillOval(ctr_x-(dx/2), ctr_y-(dy/2), dx, dy);
   break;
}
```

Note the first line of this code excerpt. The variable *val* is the attribute value being manipulated, and we are now deciding how big to draw the marker in the cell. Taking the square root of the attribute value makes the area of the dot proportional to the value of the attribute. This makes the markers more accurately represent the attribute than if the radius was proportional to the attribute.

Also note that this graph has two possible styles, *JHintonGraph.STYLE_SQUARE* and *JHintonGraph.STYLE_DOT*. These may be specified using the method *JGraphPanel.setStyle()*.

Confusion

aip.app.confusion.Confusion This project is defined in the *Confusion.jcp* project.

The test program *Confusion* for *JHintonGraph* sets up two *JHintonGraph* objects and then feeds them the data from Tables 1-7 and 1-8, respectively.

A summary of *Confusion* is listed in Table 1-10. This application is so much like *Signal* that we will only give it a brief look.

Table 1-10

Confusion summary

Public Methods	
	Confusion ()
void	assemble ()
void	run ()
Static Public Methods	
void	main (String args[])

assemble() The bulk of this method creates the graphs and sets them up.

```
m_graph1 = new JHintonGraph(300, 300);
m_graph2 = new JHintonGraph(300, 300);

Border border = BorderFactory.createBevelBorder
   (BevelBorder.LOWERED);
m_graph1.setBorder(border);
m_graph2.setBorder(border);

m_graph1.addLayer("Row Normal", Color.black);
m_graph2.addLayer("Table Normal", Color.blue);

m_graph1.setRange(0, 0, 3);
m_graph1.setRange(1, 0, 3);

m_graph2.setRange(0, 0, 3);
m_graph2.setRange(1, 0, 3);

JPanel panel = new JPanel(new FlowLayout());
content.add( panel, BorderLayout.WEST );

panel.add(m_graph1);
panel.add(m_graph2);
```

Layer 0 and layer 1 of both graphs are defined as 4 cells wide and 4 cells tall, with each cell identified in *X* and *Y* by a zero-based index in the range [0 .. 3].

run() The data is hard-coded in arrays the *run()* method, but could be read from a disk, generated from some other class, or transmitted psychically by Martians.

And Beyond

There are many statistical techniques not touched on here, from regression, to chi-squared analysis, ANOVA, factorial ANOVA, MANOVA, and so forth. If you are interested in analysis, your next stop is a good statistics book or two.

CHAPTER 2

Computing Framework

This chapter explores several architectures used by artificial intelligence programs, paying special attention to agent architectures. It then introduces distributed and network computing with threads, pipes, sockets, and the Java *ClassLoader*.

AI does not exist in a vacuum, but is created in the context of a problem and it operates within an extensive bed of other software. This could be said of any algorithm, from the lowliest bubble sort up to the most complex expert system.

While you might not normally worry about how to integrate your Euclidian parametric solver into the application, AI algorithms often need more consideration. Some AI algorithms are extremely CPU expensive and as such may run on a different machine, or several different machines, than the user is on. Software agents, by their very nature, tend to operate with or on different machines across a network of systems. Or perhaps you want the AI to operate in a separate thread from the main task, thinking in the background.

My time in AI has been split between computer-aided manufacturing applications and mobile robotics, so many of my examples reflects this history.

Architecture Overview

This section considers the AI system as a type of agent acting in an environment. For some systems, such as software agents, robotics, and game AI, this may be a fairly literal model. In other systems where the AI is simply a search, classification, or optimization system of some kind, this is a figurative example. The sense and events would map to inputs and outputs of the system, and the environment is simply the software package that the subsystem is a part of. This agent/environment black box model is shown in Figure 2-1.

There are many different ways to fill in the agent box in Figure 2-1. For example, a fairly detailed discussion in the context of mobile robotics can be found in Dudek (2000) and Arkin (1998).

Note that the various architectures below are not necessarily mutually exclusive. Sometimes hybrid systems are the best choice.

Sense–Plan–Act The classical sense–plan–act control cycle addresses a problem one step at a time, as shown in Figure 2-2.

First, sensory information is received and processed by a perception module.

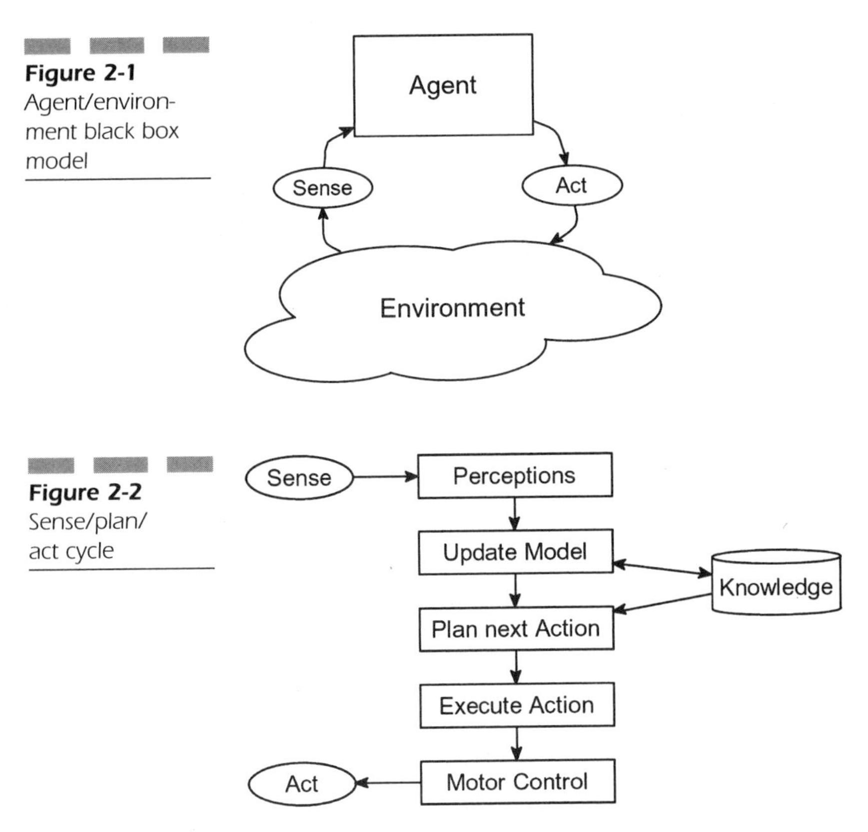

Figure 2-1 Agent/environment black box model

Figure 2-2 Sense/plan/act cycle

This in turn passes its results to the world model, which updates its knowledge of the world based on these perceptions.

From this updated model, the planner plans the next action to perform.

This plan is then passed off to an execution sub-system, which in turn drives the motor controllers to execute the plan.

This type of serial control architecture could be broken down into any number of functional units with different tasks than listed in Figure 2-2.

Classical AI is based on this architecture, and the various pieces tend to be written in symbolic languages like Lisp, or in logic programming languages like Prolog.

For mobile robotics, this control model tends to fail in the face of real-world chaos, though for better-behaved application areas it should prove more tractable.

Reactive Control The reactive control model is in contrast to the logical, thinking approach that sense/plan/act embodies. Where sense/plan/act systems carefully model their environment and then build and execute plans based on that model, reactive control systems are reflexive in nature, responding directly to stimulus. This allows reactive systems to respond rapidly to a dynamically changing environment.

As you can see in Figure 2-3, reactive systems are parallel in nature, with each individual sub-behavior receiving stimulus and sending control signals. These signals can be summed, averaged, or otherwise blended together to create the final behavior.

Subsumption is the most famous reactive control architecture. It is a technique developed by Brooks (1985) to provide layers of behavior. Higher level behaviors can override, or subsume, lower level behaviors when they need to.

Subsumption behaviors are developed in layers, or levels of competence, with higher layers implying more specific behaviors. Each layer is, in turn, composed of one or more modules. These modules were developed as augmented finite state machines (AFSM), but the implementation is not as important as the concept they embody. Each module takes the form shown in Figure 2-4.

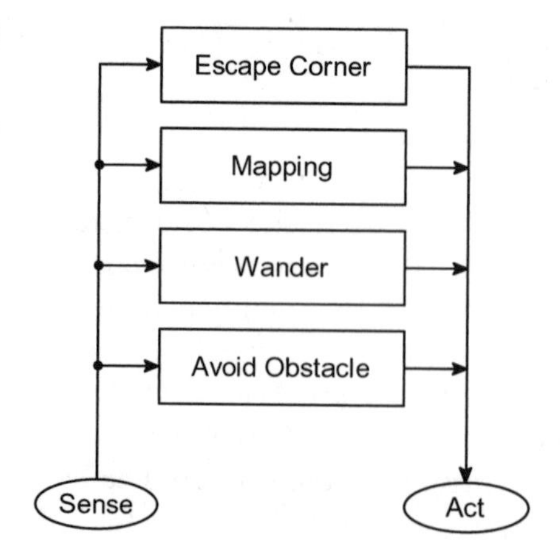

Figure 2-3 Reactive control

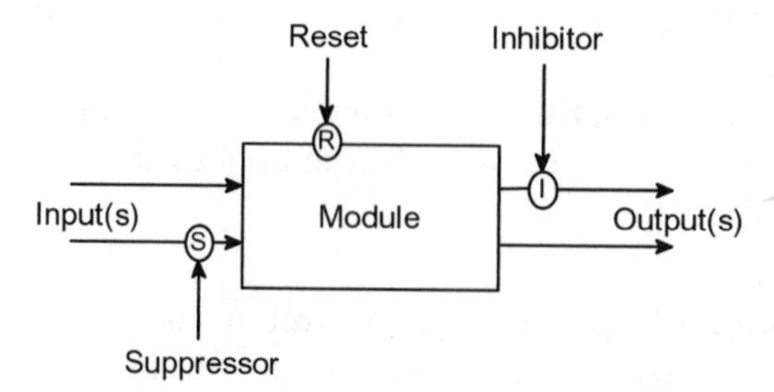

Figure 2-4 Subsumption module

A basic control module will receive one or more inputs and can in turn send out one or more outputs. The form of these signals depends on the application area you are working in.

The output of one module can be used to suppress an input to another. When this occurs, the suppressing input replaces the original input.

The output of one module can be used to inhibit the output from another. The inhibiting output replaces the original output.

Finally, a module may be reset.

A sample subsumption system from Brooks (1985) is shown in Figure 2-5. Note that there is no internal model of the world. These architectures assume the world is its own best model, and operate directly from that.

Extensions to the basic subsumption module include special timer modules and the availability of global variables (hormones) that can moderate the behavior of multiple modules.

Continuous Control Continuous control systems generate a control signal based on the current goal and feedback from the system. While they do not look like much in the black-box model (Figure 2-6), they can

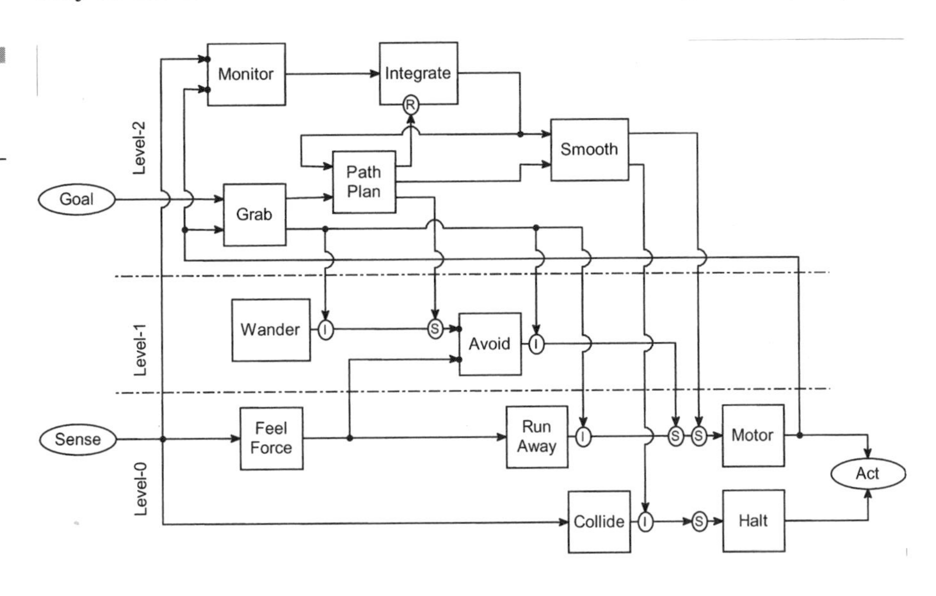

Figure 2-5 Subsumption example

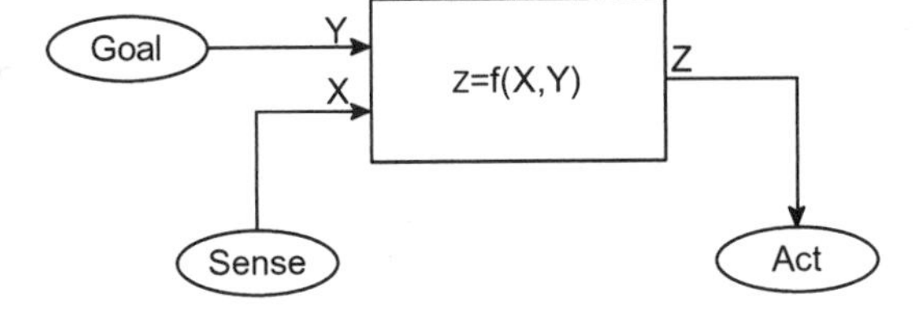

Figure 2-6 Continuous control

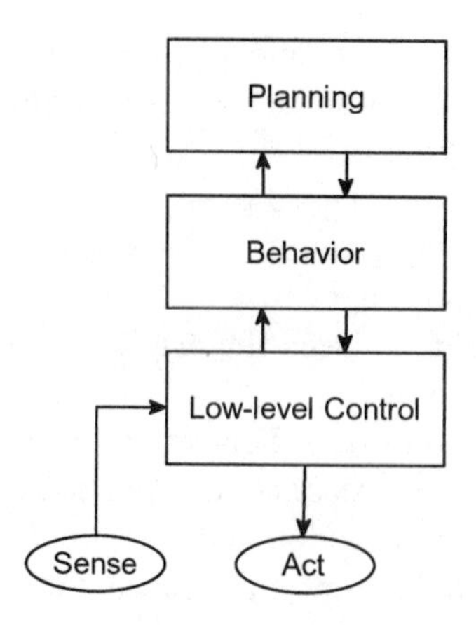

Figure 2-7 Hierarchical control

actually be quite complex. Examples of continuous control include proportional-integral-derivative (PID) control and fuzzy logic.

Hierarchical Control Hierarchical control adds depth to an otherwise flat control scheme. Figure 2-7 shows a sketch of what this could mean. At the lowest level would be the low-level control, which may be the only part of the agent that knows about motors, sensors, file systems, HTML, or whatever it is the environment presents to it.

The low-level control abstracts and buffers this information so it can be better understood by higher level behaviors. These behaviors provide sequencing and other operational directives to the low-level control.

Above it all lays the planning system, which does all of the long-range thinking and sets the behaviors in motion.

Hierarchical control looks a lot like sense/plan/act turned on its side. However, while sense/plan/act is by nature serial in operation, hierarchical control implies a parallel approach with each layer operating pretty much independently of the other.

Subsumption is a hierarchical reactive control, for example.

Blackboard Communication A blackboard is a central repository of information that is used to communicate between several computing modules (Figure 2-8). Each module is an expert in one part of the problem space. The blackboard can contain input information, partial answers, and final answers—the inputs and outputs of the expert modules.

Each expert watches the blackboard, waiting for the information they need to appear on it. When it does, that expert performs its calculation and puts the results back onto the blackboard.

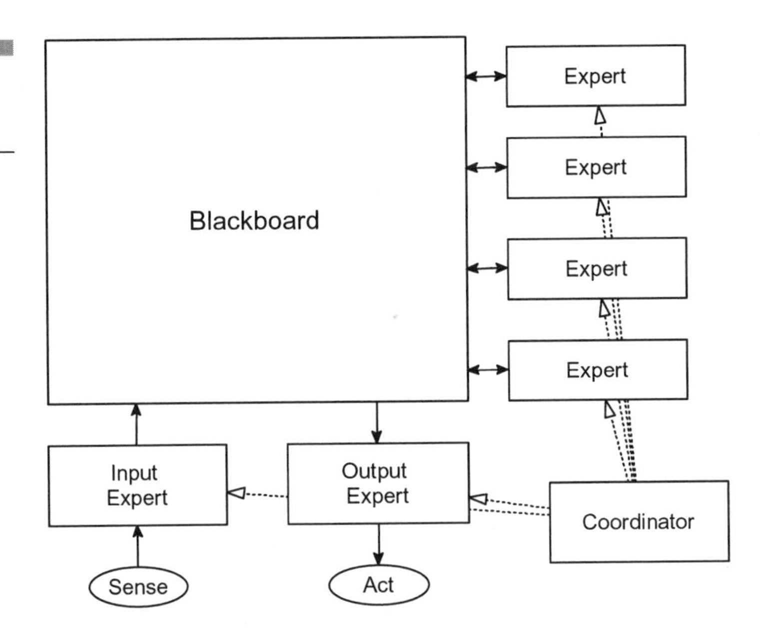

Figure 2-8 Blackboard communication

There may also be a coordinator or control module that manages the unruly herd of experts, to keep their results and calculations progressing in an orderly fashion.

More information about blackboard systems can be found in Corkill (1991) and Carver (1992).

Of course, the data store and experts that operate on it may all be on different computers. The blackboard then becomes a way to coordinate parallel computing operations. Yale's LINDA language is one environment that supports this (Linda, 2000).

Agents, Swarms, and Cellular Automata A software agent is like a virtual robot—it is a software entity that has goals, memory, and some form of intelligence. Agents interact with their environment and other agents, sometimes modifying the environment as well as each other.

You may have one agent that you send out into the world to perform its task, a group of agents that work together for a common cause, or an army of agents rampaging across the virtual terrain. Well, maybe not rampaging.

Swarm techniques use many simple processes working together, in parallel or simulated parallel, to perform a calculation or task. They are not as complex as agents, but fulfill the same idea of cooperative parallel computation.

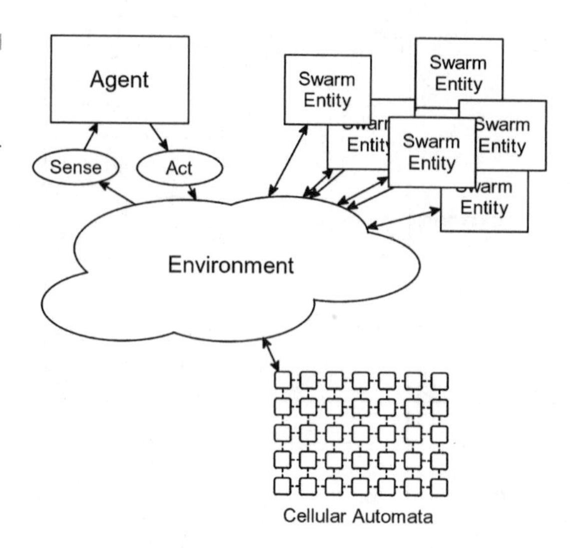

Figure 2-9 Agents, swarms, and CA

Cellular automata (CA) take the model of mass processing down to an even finer grain. CA tend to live in regular grids and perform very simple operations. Sometimes the CA grid is the environment itself, and the computation occurs by having the problem dropped into this environment, like yeast eating sugar to create alcohol.

These three "group" architectures are illustrated in Figure 2-9.

Software Agents

Software agents are growing in popularity, especially as a way to operate on the internet. Because of their growing popularity and their relevance to modern environments, software agent architectures deserve a closer look.

In a way, software agents are like robots without all the expensive hardware. For an introduction to social agents embodied in robot hardware, see Dautenhahn (1999).

Like robots, agents are independent computing systems situated in an uncontrolled, sometimes uncooperative, environment. They need to autonomously explore, understand, and manipulate this environment to achieve their goals. And, when necessary, they can interact socially in groups.

The agent model of programming is an outgrowth of the trend towards network computing and the subsequent spread of data and processing resources across multiple machines. Agent programming concepts can help tame the complexity of these modern environments.

While any mechanism that functions to achieve a goal independent of human interaction could be considered an "agent", we are not likely to think of the thermostat on the wall as an agent any more than we are likely to consider it a robot. Technically speaking, a thermostat could be considered both an agent for temperature management and a robot that operates in the domain of your central weather system.

When you say *agent* you normally mean something with a little more complexity, a little more independence. The real value of creating computer programs that exist as independent agents begins when the agent is assigned to a complex task, such as operating a space probe or planetary explorer between instructions from its Earth-bound controllers.

Or closer to home, agents can negotiate a good price on your next airplane ticket, or scour the Internet for information, and you do not have to wait around staring at the progress bar all day.

What separates the thermostat from the web spider is the degree of intelligence. Of course, software intelligence is the subject of this book, and there are many different kinds of intelligence. For the purpose of software agents, however, an intelligent agent needs to have these attributes (Weiss, 1999):

- Intelligent agents must be able to sense and react to their environment. They are not wind-up systems that blindly operate within rigid, laboratory conditions.
- They need to be self-directed, taking the initiative to achieve their objective.
- Intelligent agents can interact with other agents and, possibly, humans to achieve their goals. They may have to negotiate and cooperate with other agents, a host environment, or other people during the course of their operation.

Software agents can be created using any of the robotics or AI architectures listed above, another excellent review of which is given in Weiss (1999). There are also a number of environments and languages that are targeted directly at software agent development.

Two of these are discussed briefly here, the Open Agent Architecture (Martin, 1999) from SRI International, and the FIPA standard (FIPA, 2002). From these, you can get a sense of what agent architectures are trying to do.

Open Agent Architecture The Open Agent Architecture (OAA) was developed for internal use at SRI International and has since been opened up for more general use, under a "community license". This license enables the free use of OAA for non-commercial use.

The OAA was created to support the interoperation and cooperation of multiple agents. Agents under the OAA communicate through a specific interagent communication language (ICL). Agents are connected to each other by way of facilitators, which act as "docking ports". Note that the ICL is known in other environments as the agent communication language (ACL). The meaning is the same, though the acronyms and content will be different.

The ICL under the OAA is an extension of the Prolog language. The OAA libraries provide support for constructing, parsing, and manipulating ICL expressions. With these supporting libraries OAA agents may be written in any of the common languages, such as C, C++, Visual Basic, and Java. And, of course, Prolog.

The facilitator is an agent server and it serves to coordinate the actions of different agents. It may also act as a central data store, or blackboard, which agents can use to communicate.

When an agent registers with a facilitator it can declare a set of services it provides, expressed in ICL. These service declarations are also known as "solvables". These services can then be used, via the facilitator, to resolve a matching request from a different agent. Of course, an agent may request actions or information from the facilitator using the ICL.

None of the agents involved in a transaction need to know about any of the other agents. The facilitator provides a transparent delegation service, shuttling requests and information between agents as needed.

Since most of the agent communication is in ICL, the facilitator can eavesdrop on the process and optimize or adjust its services to improve the agent transactions.

In addition to requesting data and services, an agent may install autonomous triggers with the facilitator. A trigger is a time-delay request that is executed only when certain pre-defined conditions are met. There are four types of triggers available to the agent:

- Communication triggers allow an agent to tap into the message stream and receive copies of messages to or from the facilitator.
- Data triggers watch the information repository in the facilitator and can send messages when certain conditions there are met.
- Task triggers are arbitrary ICL goals that can trigger a message when the conditions in the goal are met.
- Time triggers provide one-shot or recurring messages based on the passage of time.

The OAA has been used to develop a number of interesting applications, including a virtual office that you can access and control remotely, multimodal maps, various speech-recognition applications, and even mobile robot control.

Foundation for Intelligent Physical Agents The Foundation for Intelligent Physical Agents (FIPA) is an international organization, based in Geneva, Switzerland, that promotes a standard for agent interoperability. Founded in 1997, FIPA has developed an impressive set of standards for agents and agent environments. In addition to these established standards, they have a number of preliminary and experimental specifications passing through their document hierarchy.

FIPA's standard is focused on facilitating agent communication. They address how messages are transferred between agents, the representation of those messages, and additional behavior such as encryption or authentication of messages.

To manage the agent communication, a FIPA-based system needs to support not only a message transport protocol to move the messages around but also a directory of services that are available to the agents.

The messages themselves are in the form of a specific ACL. A message is self-descriptive, declaring the purpose of the message, also called its performative, the content language, encoding, and meaning, as well as the content itself. The FIPA ACL is flexible, allowing for different types of message content codings.

A FIPA agent facilitator presents a directory of both services and other agents for use by an agent. Of course, the facilitator also acts as a message transport service between agents.

When an agent registers itself with a facilitator, it gets added to the agent directory service and hooks up to a message transport service. It may then advertise any services it provides in the agent directory.

Once in place an agent may use the directory services to find another agent that can provide needed services. It may then initiate a conversation with that agent.

The FIPA standards are very detailed, and seem to be growing in their acceptance. A number of environments and toolsets have been developed around the FIPA standard. A listing of these can be found on the FIPA website at `www.fipa.org`.

As environments based on FIPA such as FIPA-OS (fipa-os.sourceforge.net/) grow in popularity, more agents will be found in the wild that interact through them. These agents can then provide interesting and complex services to any other agent that works in the FIPA environment.

Distributed Computing

Distributed computing is the act of running different parts of your program on different computers. It is a step down from agents in complexity, in that there is less need for the complex message and directory services that the agent facilitators provide. The modules in a distributed application are also more tightly coupled and less independent than agents. The decision to run your program on multiple processors or not, or even as multiple threads on one processor, is an important architectural decision. It can be difficult to change the application's architecture in mid-development.

While we looked at software agents in order to provide a mental context, this section on distributed computing is practical. And, since Java's distributing computing technology leans heavily on networking, we must first explore the interface to the world beyond our computer. This will be a whirlwind tour, but if you want more details there are many books on the subject, such as Farley (1998).

Networking

"Networking" is a term that covers a broad range of technologies. The 7-Layer Open System Interconnection (OSI) model shown in Figure 2-10 is the accepted networking model, so let us start there.

The bottom three layers do the dirty work with regards to moving the data between machines. The top three layers are mostly theoretical and live inside your application. The middle transport layer sits squarely in the operating system and does a lot of the structural work packetizing and reconstructing the network data.

1. Physical Layer The physical layer deals with the actual data bits. The network cable is what connects the two physical layers together.

2. Data Link Layer The data link layer organizes packets of data into blocks of raw bits (frames) for transmission, and in the other direction it organizes raw data back into packets. The data link also handles error detection and correction, including adding checksums to the data and requesting the re-transmission of damaged data.

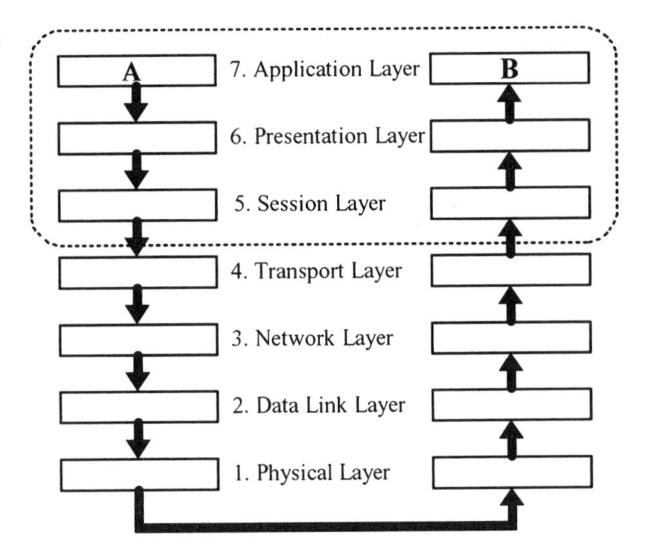

Figure 2-10 7-Layer OSI model

3. Network Layer The network layer deals with data packets. If necessary, the network layer may break a packet down into smaller packets, as needed by the lower layers. The network layer is also responsible for finding the recipient for data transmissions (addressing) and setting up the route the data will take. Internet protocol (IP) is defined at the network layer.

4. Transport Layer The transport layer works to ensure reliable service across the network. It organizes the data stream into packets and reassembles received packets back into a data stream. Universal, or "unreliable", datagram protocol (UDP) and transport control protocol (TCP) are defined at the transport layer. The commonly used TCP/IP is the TCP transport layer paired with the IP network layer.

5-7. Session through Application Layers These top three layers are more theoretical in nature, and relate to how your application communicates with the transport layer.

The session layer is an enhanced form of the transport layer. In the internet suite of protocols the sessions layer is not used.

The presentation layer defines the form of the data, in terms of records, fields, and so forth.

The application layer is everything else! Applications include FTP, DNS, Telnet, and any programs you write from this chapter.

Network Addressing In Figure 2-10, there are two applications working together through a network connection. At some point in time, these applications were not connected. The first thing that the connecting application, called the "client", had to do was to find the other application, the "server". Like mailboxes, every server application that is listening on a network has a unique address. Where people can be located by an address consisting of State, City, Street, Street Number, Apartment Number, and Name, the most common network addressing scheme is the IP.

In IP, every network interface has a unique number, like the street address of a house. On one interface there may be any number of different applications wanting to use the network connection. Applications on an interface are identified by giving each one its own socket to listen to, like an apartment number at that street address.

In its most widespread version, version 4, an IP address is a set of four numbers each of which may have a value from 0 to 255. For example, "192.168.1.1" and "67.67.213.233". Several values for the first number have special meaning, and are not used for general addressing purposes:

- 0 is a wildcard, and means "unknown".
- 127 is a "loopback" address, for when a network card is talking to itself.
- 255 is a broadcast address that will send to every network interface on the local network.
- 10, 172, and 192 are all reserved for use as a local network address and are not "public" addresses.

On my network, 192.168.1.1 is the address of my router. If you had an application that looked up that address, it would point to *your* router, if you had one, or some other network interface local to your network. Another magic address is the loopback address of 127.0.0.1, that talks to itself.

The four-number address space is divided up into five separate ranges, or address classes. Within each class, the first few bits are the class identifier, the next part of the IP address identifies the network, and the rest identifies a host interface within that network (Figure 2-11).

- Class A networks have addresses from 1.0.0.0 through 126.0.0.0. Only a few large organizations have addresses in the Class A range. Each Class A network may have 16,777,214 hosts within it.
- Class B networks have addresses from 128.1.0.0 through 191.254.0.0, and these networks are assigned to medium-sized corporations. Each Class B network may have 65,543 hosts within it.

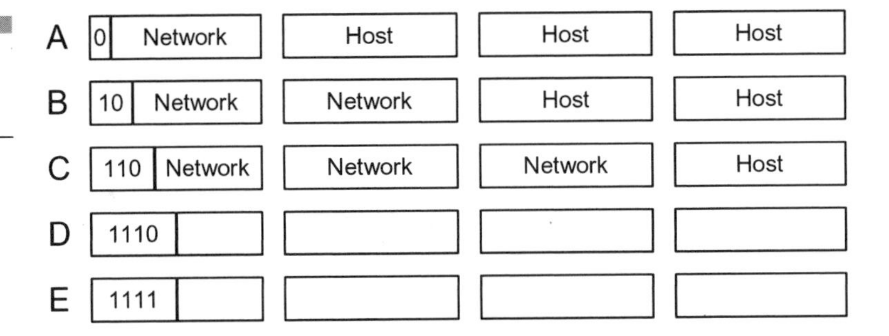

Figure 2-11 IP address format

- Class C networks have addresses from 192.0.1.0 through 223.255.254.0, and may be assigned to smaller organizations. Each Class C network may have only 245 hosts within it.
- Class D networks range from 224.0.0.0 through 239.255.255.255. These addresses are assigned to multicast groups.
- Class E networks are experimental and are reserved for future use. This class covers the remaining numbers 240.0.0.0 through 254.255.255.255.

From the core IP address there may be other considerations such as sub-network addressing, routing, gateways, and so forth, not discussed here. Of important note is the growing use of IP version 6 (IPv6), which provides for a much larger range of addresses, but with a different structure and interpretation than IPv4.

Because it is easier to remember names than arbitrary groups of numbers, there is a registry of names mapped to IP addresses. This registry is the domain name service (DNS) and most computers know the address of two or more DNS servers. These servers convert between domain names like "simreal.com" and the associated IP address "209.15.164.110". You can access your DNS server with the program *nslookup*, to look up the IP number given a name or a name given the number.

Once you have an IP address you can start thinking about the port number. Certain ports are set aside for common applications and the rest of the numbers are open for arbitrary use. Each port may be associated with the TCP or UDP protocol, or both, depending on the needs of the application listening at that port.

The port numbers are divided into three groups. Ports 0 through 1,023 are reserved as "well known" ports. Ports 1,024 through 49,151 are registered to particular users or applications, but may really be used by anyone. When you ask your computer for a free port, it will start at 1,024 and move up through this range. The remaining ports from 49,152 through

65,535 are dynamic and/or private, and are less frequently used. Some of the well-known port numbers include:

- 20, 21 File Transfer Protocol (FTP)
- 22 SSH
- 23 Telnet
- 25 Simple Mail Transfer Protocol (SMTP)
- 53 Domain Name Server (DNS)
- 80 World Wide Web (HTTP)
- 110 Post Office Protocol version 3 (POP3)
- 119 Network News Transfer Protocol (NNTP)
- 443 HTTP over SSL (HTTPS)

A complete IP address identifies the network class, the network address in that class, a host address on that network, and a port of interest on that host (Figure 2-12). For example, the simreal.com web server is at “209.15.164.110:80”. When you use a web browser the “:80” port address is implied, added on by the browser.

The IP address makes up part of the uniform resource locator (URL), discussed later under the section on Class Loaders.

Now, one interesting point is that the client and server do not have to be on different machines. They can be different programs, or program threads, running within one machine. This is where the loopback address of 127.0.0.1 can come in handy.

Which brings us to the subject of threads.

Code: Threads

Each program running on your computer is operating within a thread of execution. The operating system runs instructions from one thread for a short period of time and then jumps to another thread, on around in loops until the machine crashes or you turn it off. The rapid switching between threads gives the illusion that the different programs are all running at the same time.

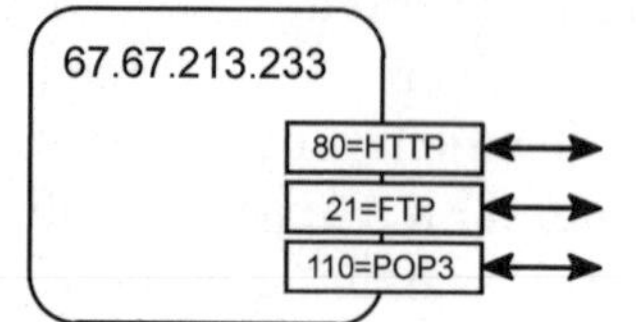

Figure 2-12 Ports on a network interface

A single program, however, may also give rise to more than one thread. Such multithreaded capability is important for servers that will have more than one client. Each client can think it has exclusive access to the server but in reality it is only talking to one of many threads within the server. Threads are also a useful tool when you want to have operations running in the background in an application.

In Java, any class may be launched as its own thread. To do this, you must define the class as either implementing the *Runnable* interface or as an extension of the *Thread* class. These approaches are roughly equivalent so use whichever is the most convenient for your program structure. Both require that you override the *run()* method with your own code.

For this example, we create two trivial objects, one as a *Thread* and the other as a *Runnable*. Each object simply performs a loop and prints the result of its own unique calculation.

Note that this trivial introductory program is not on the CD.

```
class Square extends Thread
{
   public Square(int count)
   { m_num = count; }

   public void run()
   {
       for (int idx=0; idx<m_num; idx++)
       {
           System.out.println( "Square(" + idx + ") = "+ idx*idx );
           }
       }

    private int m_num;
}

class Round implements Runnable
{
   public Round(int count)
   { m_num = count; }

   public void run()
   {
       for (int idx=0; idx<m_num; idx++)
       {
           System.out.println( "Round(" + idx + ") ="
               + Math.sin( idx / 360.0) );
       }
   }

   private int m_num;
}
```

If you called the *run()* method of each of these classes normally, the printed output would be appended, one after the other:

```
public class ThreadServer
{
    static public void main(String[] args)
    {
        ThreadServer server = new ThreadServer();
        server.serve();
    }

    public ThreadServer()
    {}

    public void serve()
    {
        Square t1 = new Square(100);
        Round t2 = new Round(100);

        t1.run();
        t2.run();
    }
}
```

```
Square(0) = 0
Square(1) = 1
Square(2) = 4
...
Square(98) = 9604
Square(99) = 9801
Round(0) = 0.0
Round(1) = 0.0027777742055340713
...
Round(98) = 0.2688724894332711
Round(99) = 0.27154693695611287
```

When these objects are launched as threads, however, the outputs are intermixed:

```
public void serve()
{
    Square t1 = new Square(100);
    Round t2 = new Round(100);

    t1.start();
    new Thread(t2).start();
}
```

Try it and see. The results are not perfectly interleaved, or even clumped into even chunks of results, but the two threads are clearly running at the "same" time.

Once you have a thread you can do a number of things to it, such as set the priority up or down with *setPriority()*, which affects how much attention the thread gets from the CPU; merge threads with *join()*; or put a thread to sleep for a period of time with *sleep()*. The Java documentation has the details for these methods and more. A good book on Java, such as Linden (1999), will also cover threads in more detail.

The threads in this example are extremely limited. You launch them and then you never hear from them again. You will normally want to send information to a thread and receive results back. This need leads us to the next topics, Pipes and then Sockets.

Code: Pipes

A pipe transports information in one direction, accepting data at its input and sending it down the pipe to its output. Pipes are used to communicate between threads of a process. Since there is no addressing system on pipes, both ends must be created and then one of the ends is sent to one of the threads. The pipe then carries information between the threads.

There are two types of pipe in Java, one for *char*-based text information, and the other for arbitrary byte data. Character pipes are composed of a *PipedInputStream* and a *PipedOutputStream*, while the byte pipe ends are *PipedReader* and *PipedWriter*. For bi-directional information flow, you need four ends defining a pair of pipes.

Let us mutate our toy program so that the values to be operated on are sent through one pipe, and the results of those operations are received through another pipe.

ThreadServer.java

aip.app.thread.ThreadServer The *ThreadServer* application is fully self-contained, including both the client and the server classes packaged nicely by the *ThreadServer.jcp* project.

A summary of *ThreadServer* is given in Table 2-1.

The action begins, as it so often does, in the *main()* method, which then activates *serve()*.

serve() *serve()* begins by creating the four pipe ends. It then turns these into two pipes using *connect()*.

Table 2-1

ThreadServer summary

ThreadServer Public Methods

```
      ThreadServer()
void  serve()
```

Square Public Methods

```
      Square (PipedInputStream in, PipedOutputStream out)
void  run ()
```

```
PipedInputStream thread_in = new PipedInputStream();
PipedOutputStream thread_out = new PipedOutputStream();

PipedInputStream server_in = new PipedInputStream();
PipedOutputStream server_out = new PipedOutputStream();
//
// Connect the ends together
//
try
{
    server_out.connect( thread_in );
    server_in.connect( thread_out );
}
```

A *Square* object is created with two pipe ends and started in its own thread with *start()*.

```
Square t1 = new Square(thread_in, thread_out);
t1.start();
```

With the *Square* off and running, we create easy-to-use Data Stream forms of the pipes, both in our main code and in the *Square* thread. *DataInputStream* and *DataOutputStream* manage all of the relevant data conversions, to and from the *byte* form.

```
DataInputStream data_in = new DataInputStream(server_in);
DataOutputStream data_out = new DataOutputStream(server_out);
```

In the main code we then drop into a loop, sending values through one pipe and receiving answers from the other pipe.

```
data_out.writeDouble( (double)idx );
data_out.flush();

double result = data_in.readDouble();
```

Note that after each *writeDouble()* call the stream is flushed. This is not necessary for the operation of the pipe, but it certainly speeds things up. Otherwise, the pipe must time out before our tiny bit of data gets sent through it.

The calls to *readDouble()* wait until enough information has come through the pipe before they return with their value.

Square(PipedInputStream in, PipedOutputStream out) *Square*'s constructor simply makes note of its two pipe ends.

run() The *run()* code mirrors the write/read calls from the *serve()* loop. Values are read from the one end of the one pipe and answers sent off in reply down the other pipe. Once the pipes are closed, the pipe read will fail and the *catch* will gracefully drop us out of the infinite loop.

```
try
{
   while (true)
   {
      double param = m_in.readDouble();
      m_out.writeDouble( param*param );
      m_out.flush();
   }
}
```

Code: Sockets

While pipes are fine when you have control of both ends of the transaction, they fall down when you need to talk between two unrelated applications. In this case, you need an addressable form of communication and this leads to sockets.

Sockets are identified by their IP address. Sockets use either the UDP or TCP protocol.

UDP is an unreliable protocol for low-priority messages, or messages that are so time-critical that with a bit of delay you might as well throw it away. Like a "wish you were here" postcard from the Bahamas. If you get home before it does, it becomes completely irrelevant. Or, more to the point, streaming audio. If one packet in your audio stream is dropped, you will get a slight glitch in the playback anyway, and getting the dropped packet *later* is really of no value; the moment has passed. The Java class for sending UDP packets is *DatagramSocket.* In the case where you may be sending to more than one receiver, or are a receiver in such a multicast group, you would use the *MulticastSocket*, which is a specialized form of *DatagramSocket*.

TCP is more reliable in that it guarantees that your packets will arrive at their destination and in their original order. Most IP applications use TCP, because it is a rare system that tolerate lost data. The Java class that sends across TCP sockets is *Socket*. The Java class that sits and listens at a TCP socket is *ServerSocket*.

Let us use these socket classes to split our threaded program into two halves. One half will be the program *SquareServer*, listening on port 50001. The other half will be a client that communicates to port 50001 to get its squares.

Table 2-2

SquareServer summary

Public Methods
SquareServer (int port)
void start ()
void stop ()
Static Public Methods
void main (String args[])
Public Attributes
final int SQUARE_SOCKET = 50001

SquareServer.java

aip.app.square.SquareServer The *SquareServer* duplicates the behavior developed for *ThreadServer* but instead of spawning a *Square* client, it listens at a port for requests.

A summary of *SquareServer* is given in Table 2-2.

main(String[] args) The *main()* entry method allocates and starts up the *SquareServer* object on the port defined by *SQUARE_SOCKET*.

SquareServer(int port) The constructor simply makes a note of the port we will be using later.

start() The *start()* method fires up a tiny window, so you can shut down the server with a click. It then creates a *ServerSocket* object, and then listens to it, waiting for a client to attach to the port.

```
   try
   { m_server = new ServerSocket(m_port); }
...
   try
   { client = m_server.accept(); }
```

Once we have a client, we create the data streams for easy communication and drop into another loop where we listen for requests and send calculations, the same as we did for pipes.

```
DataInputStream data_in =
                    new DataInputStream(client.getInputStream());
DataOutputStream data_out =
                    new DataOutputStream(client.getOutputStream());

while (true)
{
   double param = data_in.readDouble();
   data_out.writeDouble( param*param );
   data_out.flush();
}
```

stop() When the user closes the server window, the window listener *window_ear* gets the message and calls *stop()*, which closes the server socket.

```
try
{ m_server.close(); }
```

SquareClient.java

aip.app.square.SquareClient The client has to do more work than the server, though its program structure is no more complex. It looks up the server using IP addressing and then proceeds to ask the server questions.

A summay of *SquareClient* is given in Table 2-3.

main(String[] args) The client starts with the *main()* entry method as well, since it is a separate program from the server.

SquareClient(int port) Like *SquareServer*, the *SquareClient* constructor only takes note of the port number.

run() Inside the *run()* method, the client creates an *InetAddress* object that represents the IP address to connect to. Then it tries to connect to that address on the indicated port by creating a *Socket* object with that IP and port number.

```
try
{
   InetAddress ip = InetAddress.getByName( "127.0.0.1" );
   server = new Socket( ip, m_port );
}
```

The internet address used here is the loopback; we are essentially talking to ourselves. That address could be any valid IP address, however, including DNS names like "`www.simreal.com`".

Table 2-3

SquareClient summary

Public Methods	
	SquareClient (int port)
void	run ()
Static Public Methods	
void	main (String args[])
Public Attributes	
final int	SQUARE_SOCKET = 50001

If we connect to the server, *run()* drops into the familiar code that creates the data streams and queries the server through them.

Running Square Server and Client To test this system, you must first execute the server application. Then, once it is settled into place, run the client and it will report on its results.

If the server has to do any serious thinking to fulfill its duties to the client, there may be more than one client clamoring for attention at one time. In this case, the server could start up an object in a separate thread to do the thinking and return to its contemplation of the port.

Notice that once the server is up and running, *anyone* who can see this port on your network interface can connect to it and make it jump through hoops. This blind trust in your fellow programmer is rarely a good thing, but to limit access requires delving into the world of security, authorization, encryption, and signature keys. This is far outside the scope of this book. Before you start launching applications that tie in to the internet, you would be well advised to spend some quality time reading up on security issues.

Code: Class Loader

So far we have only looked at threading and networking, leading up to the topic of distributed computing. One way to distribute your computing was seen in the previous section on sockets, where a client and server passed messages to each other.

Another way to distribute your computing is to transfer a program across the network and run it on a remote machine. This requires two technologies. The first is the ability to load java code from a file and instantiate it as a runnable object, and the second is the ability to do this across the network. The concepts discussed here are also implemented in other languages, such as C#, though the programming details will be different.

Loading Arbitrary Code A Java program consists of one or more classes, each of which is compiled down into a data file that is the set of instructions that make up that class. These instructions are data that instruct the Java runtime environment (JRE) on how to behave. The JRE loads most of these class files automatically, as a byproduct of running your Java program.

The *ClassLoader* class opens up this ability to any application.

Say that you have a class that implements some useful interface. For this example, we write a "Hello, World" application that implements the *Runnable* interface, which is about as simple a class as we could possibly create.

aip.app.helloworld.HelloWorld

```
package aip.app.helloworld;

public class HelloWorld implements Runnable
{
   public HelloWorld()
   {}

      public void run()
   {
      System.out.println( "Hello, World!" );
   }
}
```

Once the file *HellowWorld.java* that contains this code is compiled, there will be a class file named *HelloWorld.class* in the *aip/app/* subdirectory.

HelloLoader.java

aip.app.helloloader.HelloLoader

We can now write a simple application to load, instantiate, and run this code. For production code, you would want to be a bit more flexible in your class identification, but for our example, everything is hard coded.

The summary for *HelloLoader* is given in Table 2-4.

load() The work in *load()* is done via the system *ClassLoader*, which is part of the *Java* system.

```
ClassLoader cl = ClassLoader.getSystemClassLoader();
```

We then instruct our newly created *ClassLoader* to load the class "`aip.app.helloworld.HelloWorld`" using its *loadClass()* method.

If the loader can find the class, we create and assign a *Runnable* object to *m_module* from the newly loaded class using *newInstance()*. Since *HelloWorld* implements the *Runnable* interface, this cast succeeds and our work is through.

```
try
{
   Class run_class = cl.loadClass( "aip.app.HelloWorld" );
   m_module = (Runnable)run_class.newInstance();
}
```

Table 2-4

HelloLoader summary

Public Methods	
	HelloLoader ()
void	load ()
void	run ()

run() *run()* simply executes the *run()* method on our previously created object in *m_module*.

Running HelloLoader If everything is set up correctly this *HelloLoader* program will load the *HelloWorld* code (which is not, in any way, linked or otherwise tied to the *HelloLoader* program) and run it.

HelloURLLoader.java aip.app.hellourlloader.HelloURLLoader

An URL is a way of identifying documents on the internet. Part of the URL is an IP address. The rest of the URL identifies the protocol to use and the document location and name:

```
<protocol>://<user>:<password>@:<host>:<port>/<path>
```

where:

- <protocol> is the specific communication protocol to use. These include:
  - ftp File Transfer Protocol
  - http Hypertext Transfer Protocol
  - mailto Electronic mail address
  - file Local file
- <user>is an optional user name
- <password>is an optional password.
- <host>is the IP address we are connecting to.
- <port>is the optional port number to connect to. The port is normally implied by the protocol used, but may be overridden as needed.
- <path>identifies the location and, possibly, document to access.

A typical URL for HTTP access to a document would look like:

```
http://www.simreal.com/index.html
```

Note that optional *user:password* and the *:port* identifier are all left off of this example.

A minor adjustment makes the previous *HelloLoader* program load its class from the internet instead of from your harddrive.

load() The only significant difference between *HelloLoader* and this *HelloURLLoader* is in the class loader. The URL loader first defines a list of URLs that it will search for classes, in this case simply the one jar file "`http://www.simreal.com/HellowWorld.jar`". It then creates an *URLClass Loader* using that list.

```
URL url_list[] = new URL[]
{
   new URL( "http://www.simreal.com/HelloWorld.jar" )
};

ucl = new URLClassLoader( url_list,
      ClassLoader.getSystemClassLoader() );
```

Essentially, this is all there is to it! If you want to try the *HelloLoader* program, feel free to use the *HelloWorld.jar* at the address shown. Otherwise, you need to compile your *HelloWorld* class into a *.jar* file and put it up on a public location.

Remote Method Invocation

Skipping along the surface of our problem space, we now come to Java's remote method invocation (RMI) system. This provides a more dynamic way to spread execution out across a network than the URL class loader.

RMI is just one of a class of systems that allows you to execute objects remotely. Other systems include CORBA and DCOM, and if you are doing mixed-language development you will want to do more research into these. However, for Java-only work, RMI is the easiest to use.

The structure of an RMI system is reminiscent of the agent frameworks discussed earlier.

On a server machine you run the RMI name server *rmiregistry*. This creates a central place where Java objects may be registered for later use.

Remote objects are registered with calls to

```
java.rmi.Naming.rebind( name, object );.
```

Clients then find the objects with calls to

```
java.rmi.Naming.lookup(name);
```

Clients can use the looked-up object as if it were local, except that execution transfers to the remote instance of the object.

Running RMI is a bit involved, however the documentation from Sun covers it fairly well, as do various books like Farley (1999).

Communication Language

When using software agents or other distributed environments, you need a way to communicate information between the various agents, the agents and the environment, or the objects in a distributed implementation.

The "easy" way is to use the messages, parameters, and the other mechanisms that you use every day in programming. These are the mechanisms in play using RMI, CORBA, remote procedure call (RPC), and other remote invocation systems.

If your software is going to join a larger community, it may be necessary to use a more generic representation for information. The development of a neutral communication language also separates the message from the messenger.

Two well-defined communication languages are the Knowledge Query and Manipulation Language (KQML) and the closely related FIPA ACL. In fact, structurally, FIPA ACL is the same as KQML, though it has a different set of commands, and it assumes that content is defined in the knowledge interchange format (KIF). KIF, then, is a logic language, a variation on first-order predicate calculus (Genesereth, 1992).

KQML demonstrates the basic concepts involved, so we present a brief overview of it here. Additional information can be found in Labrou (1999) and Weiss (1999). KQML was described as a draft standard in 1993 (DARPA, 1993), with a proposed extension in 1997 (Labrou, 1997).

A KQML message, known as a performative, is written in ASCII text with a Lisp-like syntax. The actual content of the message is opaque to KQML; it does not care what the message is. KQML is like an envelope for the message, providing information about the sender, receiver, and what the receiver can expect the message to contain.

A sample message could look like:

```
(ask-one
        :sender square_client
        :receiver square_server
        :content (192)
        :reply_with value
        :ontology _raw)
```

The first word after the opening parenthesis is the type of performative. While KQML is open-ended, there is a set of standard performatives. These are divided into 11 categories:

1. Informative performatives include tell, deny, and untell.
2. Database performatives include insert, delete, delete-one, and delete-all.
3. Basic responses include error and sorry.
4. Basic query performatives include evaluate, reply, ask-if, ask-about, ask-one, ask-all, and sorry.
5. Multi-response query performatives include stream-about, stream-all, and eos.
6. Basic effector performatives include achieve and unachieved.
7. Generator performatives include standby, ready, next, rest, discard, and generator.
8. Capability-definition performatives are limited to advertise in the 1993 specification.
9. Notification performatives include subscribe and monitor.
10. Networking performatives include register, unregister, forward, broadcast, pipe, break, and transport-address.
11. Finally, facilitation performatives include broker-one, broker-all, recommend-one, recommend-all, recruit-one, and recruit-all.

Like performatives, there is a default set of reserved parameters, though these may be extended:

- :content
- :force
- :in-reply-to
- :language
- :ontology
- :receiver
- :reply-with
- :sender

If the content is the letter and KQML is the envelope, a third aspect of the communication is the low-level mechanism used to deliver it; the post office, as it were. This third level can be anything.

Using the KQML envelope, an agent can carry on a variety of different types of conversations with other agents or service providers, advertise services, and query about services.

CHAPTER 3

Control Systems

This chapter talks about ways to process signals and control external systems using reflexive techniques. We first introduce the concept of the control system as a type of reflex. Then we look at code for filtering signals and images using convolution filters.

PID controllers are introduced and used to run an oven-like simulation. Finally, a fuzzy logic system is developed and applied to the same oven.

Introduction: Reflexes

The lowly reflex lies at the bottom level of intelligence. A reflex is defined as "an automatic reaction, without volition or conscious control". In terms of computational AI a reflex provides a mapping from inputs to outputs. A reflex does not make decisions, it simply reacts. The inputs and outputs of other more intelligent systems are buffered and managed by reflexes.

Reflexes correspond to the systems of industrial control. Control processes use feedback to bring the system to a defined set-point and keep it there in spite of disturbances to the system (Figure 3-1).

There are different ways to approach a control problem. Passive control is where you design the physical system so that it "just works". While this can be hard to change, since it requires structural modifications, it is cheap and robust.

Open-loop controls issue commands to the system but do not make use of any feedback. These require a precise internal model of the system under control, as well as a well-behaved system receiving the commands.

Closed-loop control makes use of sensors to incorporate feedback into the control loop. Closed-loop control allows for uncertainty in the model as well as noise and disturbances in the system under control.

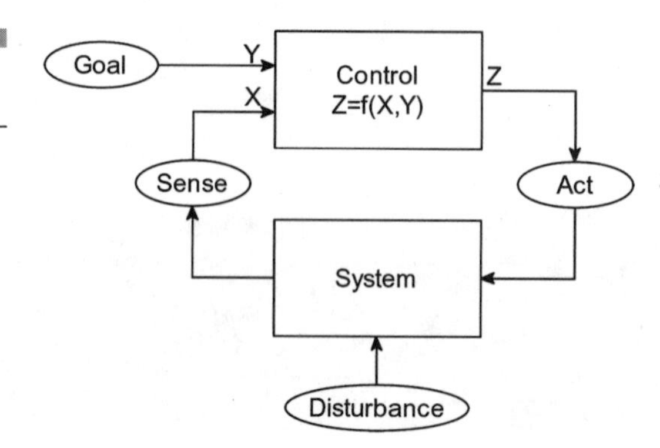

Figure 3-1
Reflextive control

The physical world, and the system we are trying to control within the physical world, is complex and often chaotic and in some cases inherently unpredictable. The field of nonlinear control gives us tools that provide stable control over these systems.

Because nonlinear systems are so complex we try to adjust them into a simpler linear form when possible. The field of linear control studies simplified, linear approximations of real-world models. Even though no real-world system is truly linear, it is still possible to control many of them using linear control processes.

While the world of control theory is outside the scope of this book, we do explore several control algorithms.

Filtering

There are two ways of thinking about filters. On the one hand a filter is like a simple open-loop control system where the output is a functional transformation of the input. On the other hand filters can be useful as signal processors used to modify the inputs, or even outputs, of other more complicated systems.

A filter takes an input signal and enhances or dampens aspects of it. For example, the equalizer on your stereo consists of several filters that change the sound of your music.

There are different types of filters. Common filters in signal processing are low-pass, high-pass, band-pass, and gap filters (Figure 3-2). We take a look at low-pass and high-pass filters here.

Electronic RC Filters

In electronics, perhaps the simplest filters of all are the RC (resistor and capacitor) filters. Figure 3-3 shows two basic RC filters.

These filters are described by their time constant τ and center frequency f_0:

$$\tau = RC$$
$$f_0 = \frac{1}{\tau 2\pi} \quad \textbf{3-1}$$

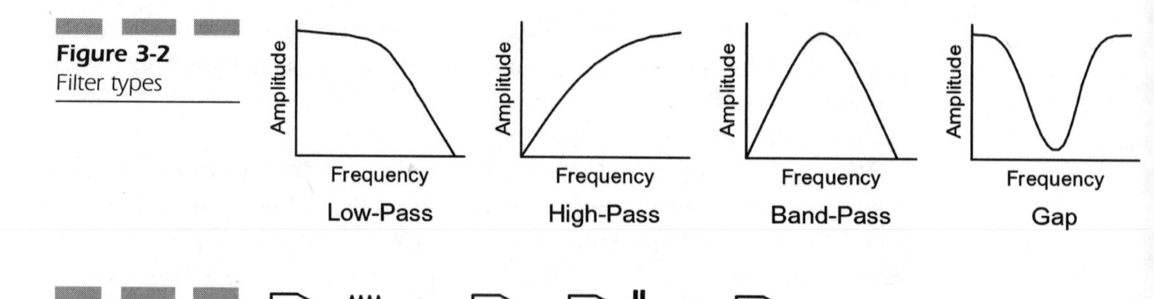

Figure 3-2
Filter types

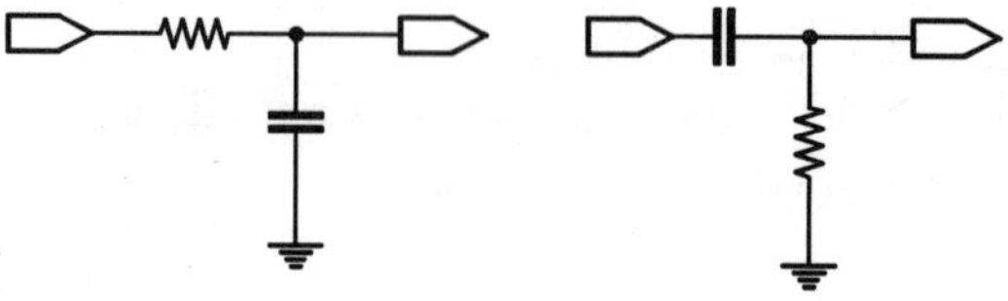

Figure 3-3
RC filters

For the low-pass filter the output amplitude factor A is defined by:

$$A = \frac{1}{\sqrt{1 + \left(\frac{f}{f_0}\right)^2}} \tag{3-2}$$

And for the high-pass filter the amplitude factor A is:

$$A = \frac{1}{\sqrt{1 + \left(\frac{f_0}{f}\right)}} \tag{3-3}$$

The filter also introduces a phase change to the signal. Of course, we are not working with electronic signals here. Our area of interest is in software filters. One simple and efficient method of filtering data is with discrete convolution.

Convolution Filters

Convolution filters are used in image and signal processing. For each datum processed, that datum and its immediate neighborhood are multiplied by the filter kernel. The neighborhood is in pixels for images or back in time for linear signals. The central datum of the neighborhood is changed by its context.

In image processing you have a large grid of values that define the image:

0,0	1,0	2,0	3,0	4,0	5,0	6,0	...
0,1	1,1	2,1	3,1	4,1	5,1	6,1	...
0,2	1,2	2,2	3,2	4,2	5,2	6,2	...
0,3	1,3	2,3	3,3	4,3	5,3	6,3	...
0,4	1,4	2,4	3,4	4,4	5,4	6,5	...
0,5	1,5	2,5	3,5	4,5	5,5	6,5	...
0,6	1,6	2,6	3,6	4,6	5,6	6,6	...
...	...	...	...	...	...	...	...

There is a picture element, or pixel, at each position in the grid with a grayscale intensity value or color information. There may be up to four intensity values, one each for the red, blue, green, and alpha (transparency) attributes of the image.

The kernel is a smaller grid of values that is multiplied against a subset of the image. For example, a 3×3 mean-value convolution kernel contains these values:

−1	−1	−1
−1	8	−1
−1	−1	−1

When you are processing an image pixel at [*x*,*y*] the coordinate mapping of a convolution kernel pixel to the image pixel is:

$x-1, y-1$	$x, y-1$	$x+1, y-1$
$x-1, y$	x, y	$x+1, y$
$x-1, y+1$	$x, y+1$	$x+1, y+1$

So the destination value *D* at [*x*,*y*] is the product of the kernel values *K* and the source image values *S*:

$$D_{x,y} = (S_{x-1,y-1}K_{0,0}) + (S_{x,y-1}K_{1,0}) + (S_{x+1,y-1}K_{2,0}) + (S_{x,y-1}K_{0,1}) + \ldots + (S_{x+1,y+1}K_{2,2})$$

The destination is created as a separate image from the source to prevent the partial results from interfering with the process.

The mean-value kernel shown above is a high-pass filter. Two other forms of high-pass filter kernels are:

0	–1	0
–1	4	–1
0	–1	0

1	–2	1
–2	4	–2
1	–2	1

A low-pass filter kernel is:

1/9	1/9	1/9
1/9	1/9	1/9
1/9	1/9	1/9

Note that the sum of all values in the kernel equals zero. If you make this sum non-zero you affect the overall intensity of the result.

Even simpler than the image filter is the one-dimensional signal filter.

Code: Convolution Filter

Many of the inputs and outputs of your control system will be one-dimensional signals. This example implements a low-pass and a high-pass convolution filter on a one-dimensional input signal. Figure 3-4 shows the program in action.

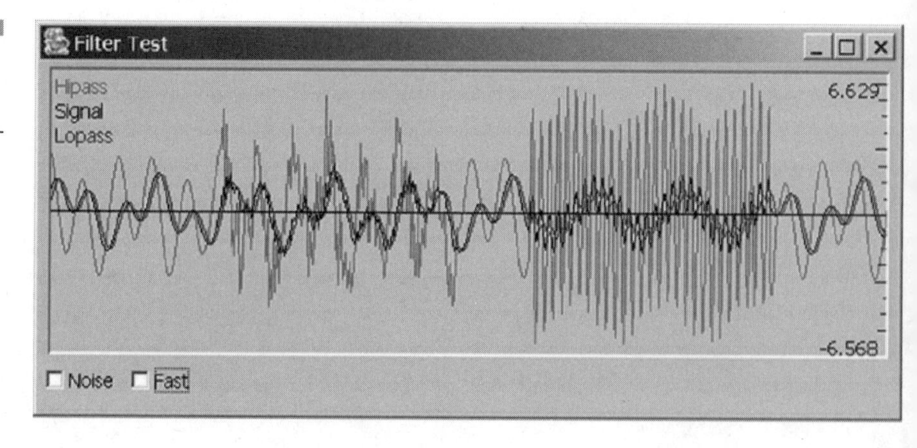

Figure 3-4 Convolution filter program

Table 3-1

Filter summary

Main Loop	
void	assemble ()
void	run ()
Filter	
void	do_filter (double signal)
double	convolve (double source[], double filter[])
Public Methods	
	Filter ()
Static Public Methods	
void	main (String args[])

Filter.java

aip.app.filter.Filter The *Filter* program demonstrates the use of a convolution filter on a time-varying signal, drawing the results in a *JLineChart*.

A summary of the *Filter* program is shown in Table 3-1.

As in most UI-based applications, most of the work is devoted to assembling the interface. This is done in the *assemble()* method. The *run()* method allocates the various filter arrays and initializes them, and then drops into an infinite loop to process the signals.

assemble() This creates the display window and constructs a three-layer *JLineChart*. One layer is for the source signal and the other two will display the high-pass and low-pass filter results.

run() During initialization, *run()* generates the two convolution filters as well as the signal history buffer *m_history*. The history buffer provides a place where we can keep a snapshot of the last few signal values for the convolution.

Once things are setup, *run()* drops into its loop. The first step of the loop generates a source signal, which is based on two sin waves. One of the waves' frequencies is based on the state of the *m_fast* checkbox. If the *m_noise* checkbox is set noise is added to the signal.

```
double v1 = Math.sin(t*10);
if (m_fast.isSelected())
{ v1 += Math.sin(t*100); }
else
{ v1 += Math.sin(t*25); }

double v2 = 0.0;
if (m_noise.isSelected())
{ v2 = Math.random() * NOISE*2.0 - NOISE; }

double v3 = v1 + v2;
```

A call to *do_filter()* performs the actual filtering and then the results are charted.

do_filter(double signal) The *do_filter()* method shifts the input signal into the signal history and then calls the *convolve()* method for both the high-pass and low-pass kernels.

```
// Shift the signal history and add this new signal
//
int num = FILTER_SIZE - 1;
for (int idx=0; idx<num; idx++)
{ m_history[idx] = m_history[idx+1]; }
m_history[num] = signal;
//
// Now set our outputs
//
m_lopass = convolve( m_history, m_loconv );
m_hipass = convolve( m_history, m_hiconv );
```

convolve(double source[], double kernel[]) The convolution is a simple process that sums the product of the history *source[]* and the filter *kernel[]*.

```
double val = 0.0;
for (int idx=0; idx<FILTER_SIZE; idx++)
{
    val += source[idx] * kernel[idx];
}
```

Running the code you may notice that while the low-pass filter does a good job of smoothing out the signal, the high-pass filter amplifies any noise. If you want to extract the high-frequency sin wave but not the noise, try running the signal through a low-pass filter to remove the noise and then the high-pass filter.

Notice that the filter size is set to five. How does the filter behave if you set the size to three? Seven? Or larger?

Instead of the calculated filters you can create filters by hand. What results do you get with different filter shapes?

Proportional-Integral-Derivative Control

Proportional-integral-derivative (PID) is a feedback control method that can be used to manage a wide range of processes. It is like cruise-control for a system; it makes sure that the process sticks to its set-point even in the face of perturbations.

PID is a three-part equation that uses the difference, or error e, of the process's current state from the commanded set-point. Each of the three parts has a different role in the controller. For simple systems one or more of the parts can be removed or their gain can be set to zero.

Using Figure 3-1 as a system model, a PID controller can be defined in different ways:

$$
\begin{aligned}
e_t &= Y - X \\
Z &= K_P\left(e_t + K_I \int e_t + K_D \frac{de_t}{dt}\right) \\
Z &= K_P\left(e_t + K_I \int e_t\right)\left(K_D \frac{de_t}{dt}\right) \\
Z &= K_P e_t + K_I \int e_t + K_D \frac{de_t}{dt}
\end{aligned}
\qquad \textbf{3-4}
$$

where:

Y is the set-point.
X is the process measurement input.
Z is the control signal output.
e_t is the error at time t.
K_P is the proportional gain.
K_I is the integration gain.
K_D is the differentiation gain.

PID controllers may be implemented in software, as firmware in microcontrollers, or as discrete electronics. They are used everywhere and there is extensive literature on their use and tuning.

In our demonstration we define a digital controller in the form of:

$$
\begin{aligned}
e_I &= e_I + e_t \\
Z_{t+1} &= K_P\,(e + K_I e_I) + K_D\,(e_t - e_{t-1})
\end{aligned}
\qquad \textbf{3-5}
$$

where e_I is the error accumulated over time.

The quality of a controller is determined by several things, such as how responsive it is to changes in the set-point, how far it overshoots a new set-point, and how long it takes to settle down to an acceptable error level. The goal is to make the process follow the set-point as closely as is reasonable, though smoothing the transitions between set-points can also be desirable.

Before we can start building a controller we need a process to control.

Code: Simplified Heater Process

There is an endless array of possible processes we could simulate, from chemical processing plants, ovens, kilns, pumps and basins, motors, and so forth. Each process has unique characteristics.

This simulation is of an unrealistically simple heating system. It is a bare-minimum process that was created for the sole purpose of demonstrating controller behaviors. There is no sensor lag or sensor noise, though this would be easy enough to add in. The other various system parameters are arbitrary.

The class is, cleverly enough, named *Heater*.

Heater.java

aip.app.heater.Heater A summary of *Heater* is given in Table 3-2.

As mentioned above, this is not meant to be a realistic simulation.

Heater(double dt) The constructor for the *Heater* class receives the time base used by the simulation. Too large a value can make it unstable, too small and it is boring. The constructor also sets the various state variables.

getState( ) *getState()* returns the current temperature, as calculated by the last call to the *Step()* method.

Step() *Step()* updates the state of the simulation. First, the control signal is clamped between 0 and 100. Then the temperature is updated based on the heater temperature and any heat lost. Finally, the heating element temperature is updated based on the control signal and heat lost. The code is almost shorter than its description.

```
if (control < 0.0)
{ control = 0.0; }
else
if (control > 100.0)
{ control = 100.0; }
//
// Update the state.
//
m_temp += m_dt*(m_heater - RADIATE*m_temp);
m_heater += m_dt*(control - RADIATE*m_heater);
```

Table 3-2

Heater summary

Public Methods	
	`Heater (double dt)`
`double`	`getState ()`
`void`	`Step (double control)`

Table 3-3

PID summary

Main Loop

```
  void  assemble ()
  void  run ()
```

Process signal generators

```
double  signal_square (double t)
double  signal_trapezoid (double t)
```

PID controller

```
double  do_pid (double err, double pgain, double igain, double
        dgain)
```

PID.java

aip.app.pid.PID The PID controller is also fairly simple, though it has the added bulk of a user interface.

Table 3-3 shows a summary of the PID test application.

assemble( ) *assemble()* creates the window and fits a wide range of graphs and input components into it. This *assemble()* is a little more interesting than previous ones due to its use of various *Box* layout styles.

run( ) The *run()* method starts by creating a *Heater* object and initializing the various gain parameters.

For each cycle of the *run()* loop a signal is generated by *signal_trapezoid()* or *signal_square()*. This signal is a time-varying set-point which the PID algorithm uses as its input.

The various gain values are grabbed from the user interface using *input_double()*. This method is "safe", in that if the user is editing a value and it becomes un-parsable a default value is returned instead of an error.

Using the current process state and set-point, we calculate the current error and get a new control value from *do_pid()*.

Finally, the heater processes are updated with this control value and the results are charted.

signal_square(double t)
signal_trapezoid(double t) The square wave signal toggles between low and high output states twice for every period.

The trapezoid is like the square wave except the cycle is divided into eight parts instead of two. Each transition between low and high takes one eighth of the cycle time to perform, giving smooth transitions.

Trapezoid or other less abrupt set points provide a better signal for the controller to follow. Rapid transitions like the square wave can cause the error integral to "wind up" and force larger overshoot.

do_pid(doiuble err, double pgain, double igain, double dgain) The math in *do_pid()* implements Equation 3-5.

```
m_ed = err - m_e;
m_e = err;
m_ei += err;

if (MathConst.isZero(igain))
{ m_ei = 0.0; }      // Avoid windup if I is off

m_out = pgain*(m_e + igain*m_ei) + (dgain*m_ed);
```

Running the PID controller Now that you have a running PID, you can explore what the various gain values do. One or more gains can be set to zero, giving you PI, PD, and P controllers, and other permutations.

The proportional gain affects how quickly the control tracks the error. This piece of the system affects the controller by paying attention to the past. Larger values allow the control signal to quickly track the error but when the gain gets too large the system can become undamped and unstable. Too small a proportional gain makes the system sluggish. Note that with P-only control there is a steady-state offset between the system state and the set-point.

The integral component is also known as the reset and it exists to remove the proportional offset. This is a sensitive gain control, with too large a value driving the controller into oscillations. One problem with the integral control is windup. Windup is when the error accumulates and then drives the control signal into an overshoot. Careful tuning of the parameters helps, as well as preventing the set-point from making large steps in value.

The derivative portion of the control tries to peer into the future, based on how the error signal is changing. While the derivative gain makes the control signal a bit snappier on the uptake, its real value comes into play when there is a delay in the sensor information, or sensor lag. This simulation does not have any sensor lag. One problem with the derivative comes when there is sensor noise. Since it amplifies the changes from step to step the sensor noise is amplified. A low-pass filter on the sensor can minimize this problem.

There are many different ways to structure PID controllers, and many different ways to tune them. Since this control technique has been around for a long time there is an abundance of resources in industrial control books and websites.

Fuzzy-Logic

A fuzzy logic system is a way to smoothly map input values to output control using a collection of if-then rules.

Fuzzy logic can also be used for decision-making and a variety of other purposes. Fuzzy logic rules, especially when written in a "natural" linguistic form, provide a way to capture human expertise in a problem without requiring the designer to translate it into abstract mathematical form.

Fuzzy logic is an extension of Boolean logic so it makes sense to look there first for our foundation.

Crisp Boolean Logic

Boolean Sets and Rules In Boolean, or "crisp", logic, a statement is true or it is false. For example, using our oven example from the previous section, if the temperature error is within 15 units of the set-point the temperature could be considered *Just Right*. However, if the oven temperature falls 15 units below the set-point it might be considered *Cool* and if for some reason it is down by 60 points or more it is *Cold*. An example of how the temperature error could be categorized is shown in Figure 3-5. This error e is described in Equation 3-6.

$$e = T_{\text{set}} - T_{\text{oven}} \qquad \textbf{3-6}$$

Note that this is an abstract problem using generic "units". For any given temperature error, from 100 units over the set-point, *Hot,* to 100 units under the set-point, *Cold*, we know exactly which category that temperature belongs to and membership is an all-or-nothing event.

Each category is defined as a set of all related temperatures. With these five temperature sets we can make up some rules of operation for our heater controller, one rule per set:

1. If e is *Hot* then turn heater *Off*.
2. If e is *Warm* then turn heater *Off*.

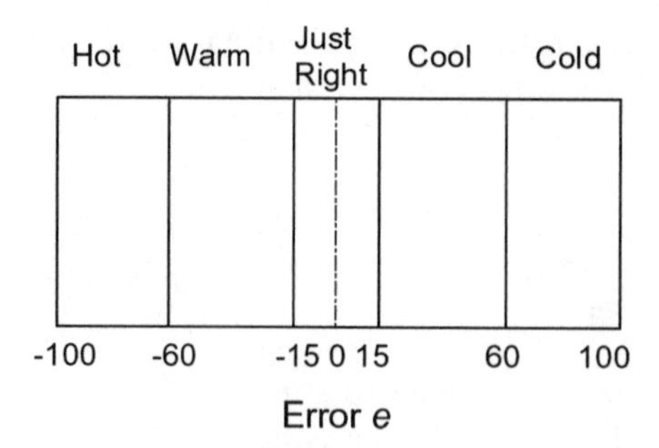

Figure 3-5
Crisp temperature evaluation

3. If *e* is *Just Right* then turn heater *Off*.
4. If *e* is *Cool* then turn heater *Low*.
5. If *e* is *Cold* then turn heater *High*.

where:

- *Off* is heater power 0.
- *Low* is heater power 25.
- *High* is heater power 100.

Note that the first three rules are the same so they could be merged. If we had control of a fan or a cooling cycle, those rules could manage the cooling cycle. This example is limited to a single input, the temperature error, and a single output, the heater power.

The first part of each rule, such as "If *e* is *Cool*" is the condition, or antecedent, for the rule. The condition must be true for its rule to become active. The second part of the rule, "then turn heater *Low*" is the output, result, or consequent of that rule. The general form for rules is:

If *X* is *A* then *Y* is *B*.

where *X* is the input variable being categorized, *A* is a particular category, *Y* is the output control variable, and *B* is a value to assign to the output. For these Boolean rules only one rule at a time is active so all of the thinking can be done with a series of *if-then-else* program statements.

```
if (m_e < 15)
{ m_out = OUT_OFF; }
else
if (m_e < 60)
{ m_out = OUT_LOW; }
else
{ m_out = OUT_HIGH; }
```

The test program *Crisp* is a duplicate of *PID* but with this crisp control logic replacing the PID math.

As defined above, the controller bounces up to the target temperature and then oscillates around the 15-unit boundary of *Just Right*, as shown in Figure 3-6.

Boolean Logic If a control system has more than one input the rules will have two or more conditions combined logically to generate an output. Logical operations include *and, or*, and *not* giving us possible rules in these forms:

If (X is A) and (Y is B) then Z is C
If (X is A) or (Y is B) then Z is C
If (X is A) and not (Y is B) then Z is C

and so forth.

The easiest way to describe the behavior of a logical operation is with truth tables like the ones shown in Table 3-4.

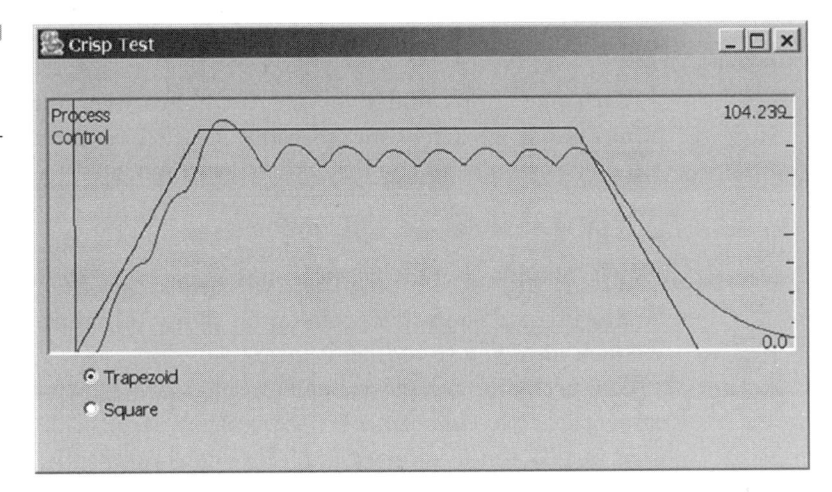

Figure 3-6 Crisp temperature controller

Table 3-4 Truth tables

OR			AND			NOT	
A	B	A or B	A	B	A and B	A	not A
F	F	F	F	F	F	F	T
F	T	T	F	T	F	T	F
T	F	T	T	F	F		
T	T	T	T	T	T		

Figure 3-7 Crisp rules with two inputs

B \ A	Very High	High	Just Right	Low	Very Low
Very High	A is VH B is VH	A is H B is VH	A is JR B is VH	A is L B is VH	A is VL B is VH
High	A is VH B is H	A is H B is H	A is JR B is H	A is L B is H	A is VL B is H
Just Right	A is VH B is JR	A is H B is JR	A is JR B is JR	A is L B is JR	A is VL B is JR
Low	A is VH B is L	A is H B is L	A is JR B is L	A is L B is L	A is VL B is L
Very Low	A is VH B is VL	A is H B is VL	A is JR B is VL	A is L B is VL	A is VL B is VL

- *(A or B)* is true if *A* is true or *B* is true.
- *(A and B)* is true if both *A* and *B* are true.
- *(not A)* is true if *A* is not true.

With one input and five sets there are five rules. If there are two inputs, each with five membership sets, the system is described by 25 intersecting rules of the form shown in Figure 3-7:

If (*X* is *A*) and (*Y* is *B*) then *Z* is *C*

If there are three inputs there can be *5*5*5 = 125* rules, etc. This combinatorial explosion is a problem for many different rule-based systems. We look at a way to reduce it later.

One problem with crisp Boolean sets is the sharp distinctions between categories. Why is it that a temperature error of 60 units is considered *Cool* but 61 units is now *Cold*? This does not really mesh well with our experience of coldness where 61 is only slightly colder than 60 and not in an entirely different category of coldness.

Fuzzy logic provides a way to operate in shades of gray, as opposed to the black and white distinctions of Boolean logic.

Basic Fuzzy

This section looks at a basic fuzzy rule system like the Boolean rules created in the previous section. After this example we explore the concept of fuzzy logic further.

Fuzzy Sets and Rules Looking at Figure 3-8 you can see that membership in these overlapping fuzzy sets is not an all-or-nothing proposition. A temperature can be, for example, somewhat *Cool* and *Cold* at the same time. As temperature error increases, the membership in *Cold* increases and membership in *Cool* decreases until it is entirely *Cold*.

Where Boolean sets have either 0 or 100% membership, Fuzzy set membership falls on a continuous scale of 0–100%. An error value of $e = 65$ might be about 30% *Cool* and about 60% *Cold*, as shown in Figure 3-9. The act of converting an input value to a membership percentage in one or more fuzzy sets is called fuzzification. The shape that controls this membership weight is the membership function.

Figures 3-8 and 3-9 show the common trapezoid function and a triangle function, which is just a degenerate trapezoid, though other shapes can also be used. Any membership shape can be approximated by a series of line segments. This is shown in Figure 3-10.

Fuzzy rules can be stated the same as Boolean rules:

1. If e is *Hot* then turn heater *Off*.
2. If e is *Warm* then turn heater *Off*.
3. If e is *Just Right* then turn heater *Off*.
4. If e is *Cool* then turn heater *Low*.
5. If e is *Cold* then turn heater *High*.

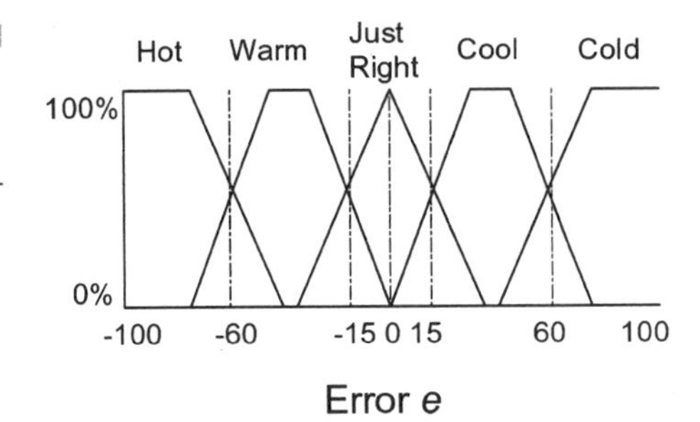

Figure 3-8 Fuzzy temperature sets

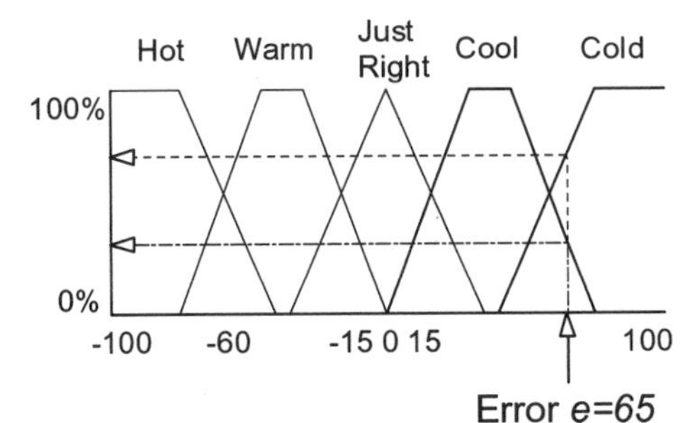

Figure 3-9 Fuzzy temperature evaluation

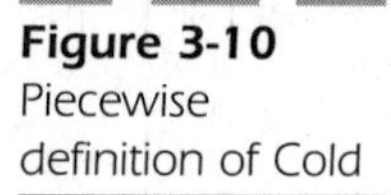

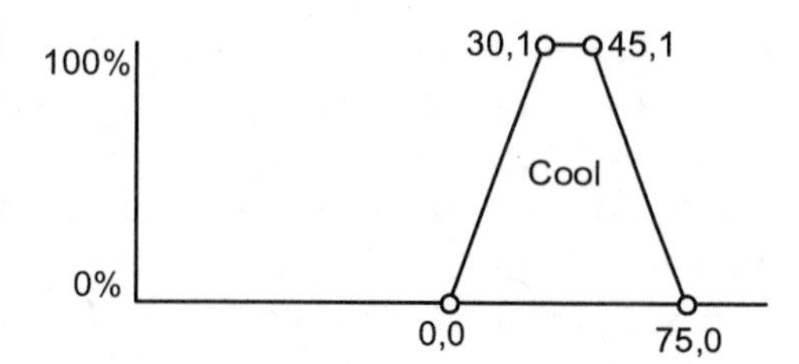

Figure 3-10 Piecewise definition of Cold

However, the interpretation of a fuzzy rule is different. In Boolean logic only one rule is active at a time. In fuzzy logic more than one rule may be active, but each to a different degree. It is necessary to blend the results of each active rule.

Each rule is activated to the extent that its condition is true. So if *e* is 30% *Cool*, the result of *turn the heater Low* is weighted at 30%. Likewise, *Cold*'s result is weighted at about 60%. These numerical results are known as singletons and they can be blended as shown in Equation 3-7. The act of blending the results of fuzzy rules into a single result value is called defuzzification.

$$R = \frac{\sum_i M_i R_i}{\sum_i M_i} \qquad \textbf{3-7}$$

where:

M_i is the membership value, or weight, of rule i

R_i is the result of rule i

The entire process is illustrated in Figure 3-11. This example, from fuzzification to defuzzification, is just one way to calculate fuzzy results.

Though we do not explore it here, the "Input Processing" box represents an important step in this, and many other, AI systems. Raw inputs to the fuzzy control process may or may not be in a useable form and may require modification before use. An example is the error value presented to the PID controller. It is calculated from the temperature and set-point, making it a pre-processed input.

Fuzzy Logic Fuzzy logic conditions can be combined and manipulated. In addition to the standard *and*, *or*, and *not*, fuzzy conditions can include intensity modifiers like *very* and *somewhat*. These intensity modifiers are discussed later.

Since fuzzy sets are not simple true/false tests but offer results along a continuum, a simple truth table does not suffice to describe them.

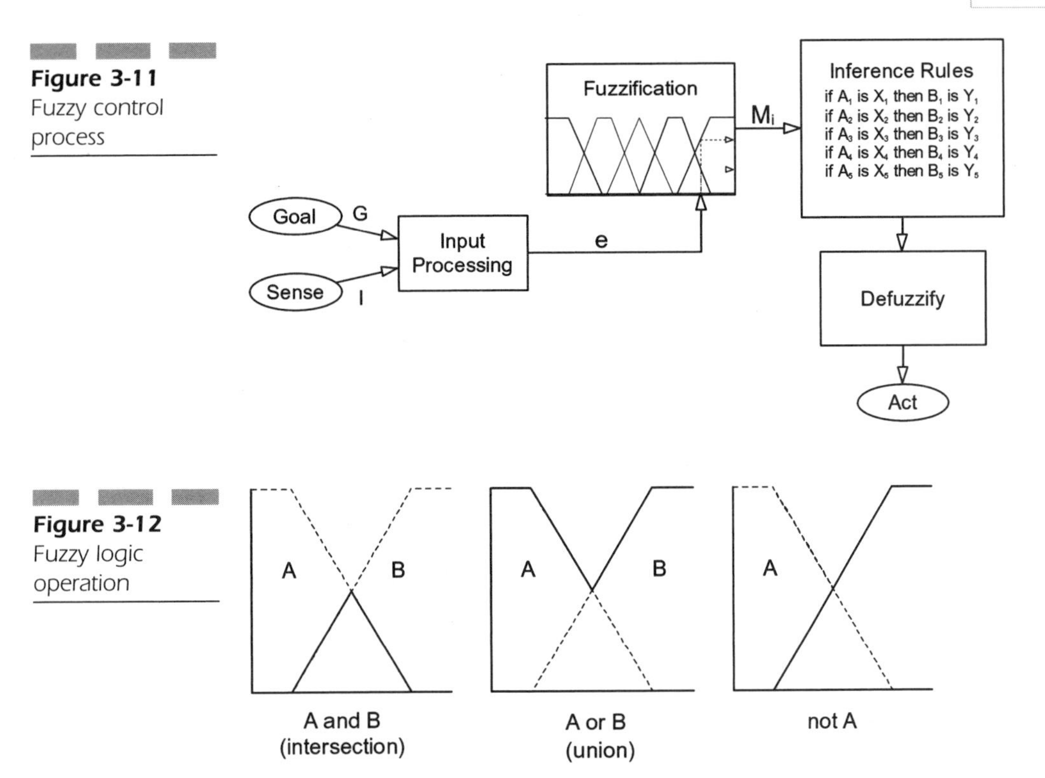

Figure 3-11 Fuzzy control process

Figure 3-12 Fuzzy logic operation

Instead, the logical operations are defined as shown in Equation 3-8 and Figure 3-12. *And* in a fuzzy set is actually the intersection of the sets, and *or* is the union.

$$
\begin{aligned}
A \cap B &= \min(A_x, B_x) \\
A \cup B &= \max(A_x, B_x) \\
\neg A &= 1 - A_x
\end{aligned}
\qquad \textbf{3-8}
$$

These are the simplest fuzzy logic definitions. There are others to be found in the literature (Kosko, 1997; Cox, 1994, and others).

Code: Singleton Fuzzy Logic

Our fuzzy heater controller will be using singleton-based fuzzy rules. A singleton is a fuzzy rule that returns a single value, like a spike. A singleton with the value of 25 is shown in Figure 3-13.

Figure 3-13 Singleton

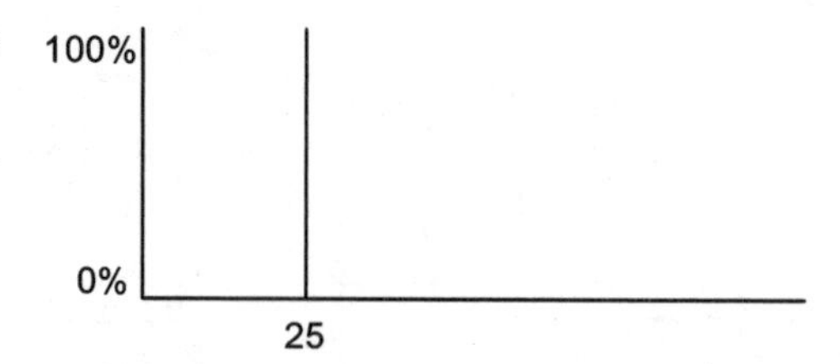

Fuzzy Number In the simplest form a fuzzy logic system can consist of a single class that represents a fuzzy number. A fuzzy number is a two-part number containing a value and a weight. The value is actually the value times the weight. When the fuzzy number is made crisp the correctly weighted result is returned.

```
public class FuzzyValue
{
    public FuzzyValue(double value, double weight)
    {
      m_value = value*weight;
      m_weight = weight;
    }

    public void add(FuzzyValue value)
    {
      m_value += value.getValue();
      m_weight += value.getWeight();
    }

    public void normalize()
    {
      m_value = defuzzy();
      m_weight = 1.0;
    }

    public double defuzzy()
    {
      return m_value / m_weight;
    }

    public double getValue()
    { return m_value; }

    public double getWeight()
    { return m_weight; }

    private double m_value;
    private double m_weight;
}
```

This code is the simplest distillation of fuzzy logic and it leaves a number of questions unanswered. The biggest question is, how is the weight determined? It could come from any number of sources, but in most fuzzy logic systems it comes from the membership functions.

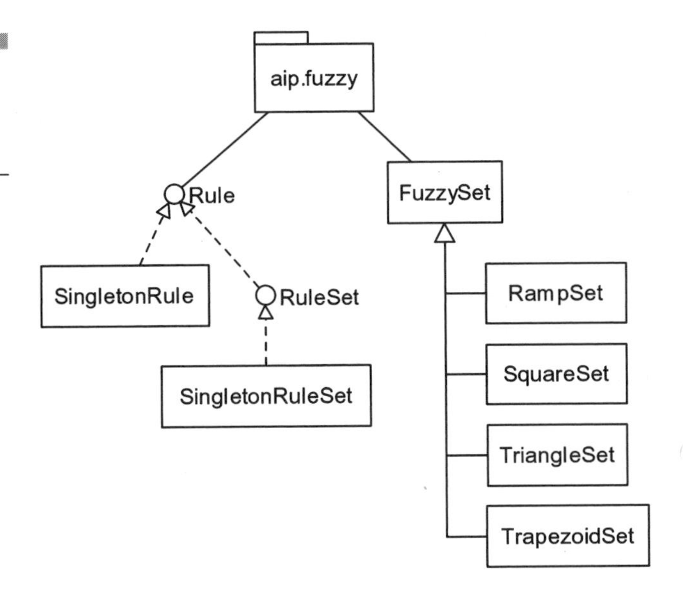

Figure 3-14 Singleton fuzzy logic class diagram

Fuzzy System Where the PID controller was, essentially, one line of math, a fuzzy logic controller requires a fairly bulky system to support it.

The classes introduced here to implement the fuzzy control process from Figure 3-11 are shown in Figure 3-14.

This is a lot of code to replace just one line of math. There are two benefits to fuzzy logic, however, not the least of which is that it applies to a wide range of problems. The other is its ability to capture expert understanding.

Fuzzy logic is at least as capable as PID controllers (Li, 2001) and in fact can be considered to be a piecewise PID control system. A fuzzy rule system can be developed to provide any arbitrary response curve or, for multi-dimensional inputs, response surface (Kosko, 1997).

FuzzySet.java

aip.fuzzy.FuzzySet The *FuzzySet* class is used to represent membership sets. These, in turn, perform the fuzzification of input values needed by the inference rules. This class represents the fuzzy membership set as a series of N points which, when strung together in order, represent $(N-1)$ straight lines. This representation is suitable for all types of membership set shapes.

Another way to represent a membership set is through some parametric function. For example, a trapezoid set can be defined by its center, plateau width, and overall width. Or a set could be a Gaussian function or other smooth curve.

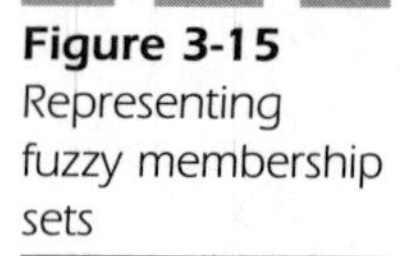

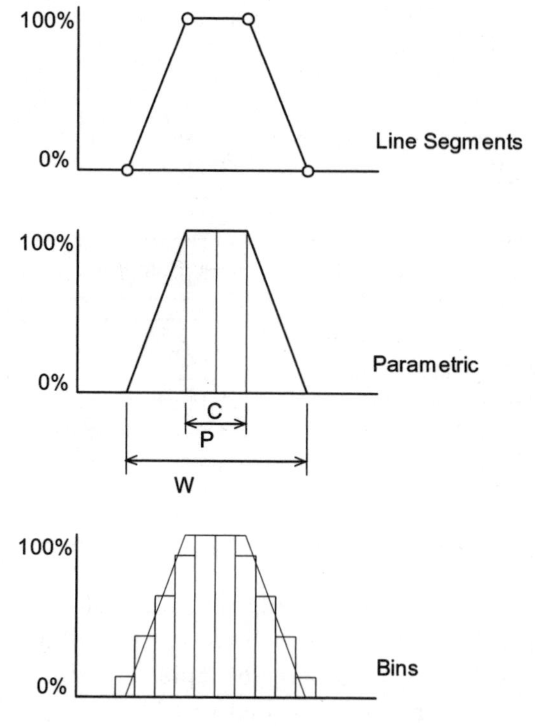

Figure 3-15 Representing fuzzy membership sets

A less precise but much faster way to define the membership set is as a series of discrete "bins" in an array of membership values `double weight[range]`. When an input value falls into the range of one of the bins the membership weight of that bin is used.

These different representation schemes are illustrated in Figure 3-15. A summary of *FuzzySet* is given in Table 3-5.

FuzzySet(int size) For efficiency the fuzzy set is represented as a fixed array of *Point2D* objects. For this to work *FuzzySet* is constructed with the final number of points it will contain. The *X* ordinate of the point represents a position along the input range and the *Y* ordinate is the membership weight associated with this input. Other positions and weights are interpolated between points to represent the entire set.

getWeight(double input) The degree of membership is calculated by finding the *Y* position in the set for a given *X* input. This is done by simple interpolation between the point whose *X* value is less than the input and the point whose *X* value is higher than the input.

Table 3-5

FuzzySet summary

```
Construction
                    FuzzySet (int size)

Calculations
          double    getWeight (double input)
          double    getCOG ()

Get/Set Access
       final int    getSize ()
 final Point2D      getPoint (int idx)
  final double      getX (int idx)
  final double      getY (int idx)
    final void      setPoint (int idx, Point2D pt)
    final void      setPoint (int idx, double x, double y)
```

```
double slope = (hipt.getY()-lopt.getY()) /
               (hipt.getX()-lopt.getX());
return lopt.getY() + slope*(input - lopt.getX());
```

getCOG() This method calculates the center of gravity, or COG, of the set. This is really a defuzzification process that would more accurately belong to a rule, however it is more convenient to put it here. Note that, in many cases, the *FuzzySet* is not only used to fuzzify input values but is used to represent output results.

While technically the COG is calculated using the formula in Equation 3-9, because of the simplicity of our piecewise "function" we can use a simpler formula.

$$\overline{x} = \frac{\sum_{i} \int_{x_{i-1}}^{x_i} xM\mathrm{d}x}{\sum_{i} \int_{x_{i-1}}^{x_i} M\mathrm{d}x} \qquad \textbf{3-9}$$

which, for our limited case, can be computed as:

```
double numer = 0.0;
double denom = 0.0;

double Xa = getX(0);
double Ya = getY(0);

for (int idx=1; idx<num; idx++)
{
    double Xb = getX(idx);
    double Yb = getY(idx);

    double dx = Xb - Xa;
```

```
        double n = dx * (Ya*(2*Xa + Xb) + Yb*(Xa + 2*Xb));
        numer += n;
        double d = dx*(Ya+Yb);
        denom += d;

        Xa = Xb;
        Ya = Yb;
    }
    return numer / (3*denom);
```

The derivation of this math can be found in Masters (1993), page 314.

get/set The get and set methods are all straightforward, accessing the various attributes of the *FuzzySet* and its points.

RampSet.java
SquareSet.java
TriangleSet.java
TrapezoidSet.java These four descendents of *FuzzySet* provide initialization shortcuts for the set's lines.

RampSet (double low, double high) A ramp set is defined by two points. The parameter *low* specifies the input value where the ramp is at zero membership. The parameter *high* is the input value where the ramp is at full membership. These *X* values may be in either order, depending on if the ramp is climbing or falling.

All inputs to the left of the ramp's lowest *X* ordinate take that ordinate's weight, and all inputs to the right of the ramp's largest *X* ordinate take that ordinate's weight.

This can be confusing, as simple as it is. For example, if *low* is set to 10 and *high* is set to 20, all inputs less than 10 have zero weight, inputs from 10 to 20 have increasing weights, and inputs from 20 upward have full membership.

On the other hand, *low* of 20 and *high* of 10 is the reverse. All inputs less than 10 have full membership, inputs from 10 to 20 have decreasing membership, and inputs from 20 on have zero membership.

SquareSet(double center, double width) While you might expect the square membership set to be defined by its start and end values, it is more consistent with the other set constructors to define it by its center and width. For input values in the range $\left[\left(\text{center}-\frac{\text{width}}{2}\right)...\left(\text{center}+\frac{\text{width}}{2}\right)\right]$ the membership weight is one and for all others the membership is zero.

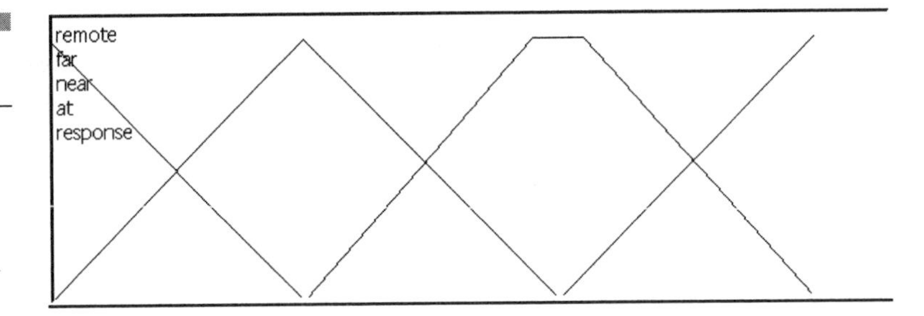

Figure 3-16 Membership sets

TriangleSet(double center, double width) A triangle set ramps from zero membership at the edge up to full membership at the center and back down to zero membership at the other edge.

TrapezoidSet(double center, double plateau, double width) A trapezoid is a triangle with a flat top. The *plateau* parameter specifies the width of this flat top, with half the width on the low side of center and the other half on the high side of center.

The ramp down, triangle, trapezoid, and ramp up sets are illustrated in Figure 3-16.

Rule.java

aip.fuzzy.Rule *Rule* provides a common interface for different types of rules. *Rule* is the leaf of a composite pattern. Its *evaluate()*, *result()*, and *defuzzy()* methods are described under the different implementations of *Rule*.

RuleSet.java

aip.fuzzy.RuleSet A *RuleSet* is a collection of one more more *Rule* objects. *RuleSet* is an extension of *Rule*, and represents a node in the composite pattern.

SingletonRule.java

aip.fuzzy.SingletonRule *SingletonRule* implements the *Rule* interface and contains a complete inference rule. The condition of the rule is a *FuzzySet* that is evaluated against the input value and the result of the rule is a *FuzzySet* that represents the output value of this rule scaled by its membership weight.

Singleton rules are special in that their result is a single point *X,Y* where *X* is the recommended output and *Y* is the weight of that output.

A summary of *SingletonRule* is given in Table 3-6.

SingletonRule(FuzzySet condition, double value) The constructor defines a *SingletonRule* with a membership function and a result value. Since rules are assumed to be normalized, with an output weight of one for full membership, the result value can be specified as a simple double with the implied weight of one.

Internally, the result is also stored as a *FuzzySet* with a single point since this is the form required by the rest of the fuzzy system.

evaluate(double input) When the rule is evaluated against an input it generates a *FuzzySet* result that has a single point with *X* equal to the result value and *Y* at the membership weight.

If the value does not fall in the range of the membership set, *evaluate()* returns *null*.

result() This method simply echoes the result of a previous *evaluate()* call.

defuzzy() Defuzzification of a singleton rule is trivial. The center of gravity is at the singleton's *X* position, so that is what is returned.

SingletonRuleSet.java

aip.fuzzy.SingletonRuleSet *SingletonRuleSet* implements the *RuleSet* interface and contains zero or more *SingletonRule* objects.

The *RuleSet* provides a way to group several rules together and evaluate them against a single input.

A summary of *SingletonRuleSet* is given in Table 3-7.

Table 3-6

SingletonRule summary

Public Methods

```
         SingletonRule (FuzzySet condition, double value)
FuzzySet evaluate(double input)
FuzzySet result ()
  double defuzzy ()
```

Table 3-7

SingletonRuleSet summary

Construction

```
         SingletonRuleSet ()
    void addRule (SingletonRule rule)
    void addRule (SingletonRuleSet ruleset)
```

Evaluation

```
FuzzySet evaluate (double value)
FuzzySet evaluate ()
FuzzySet result ()
  double defuzzy ()
```

addRule(SingletonRule rule) This adds a rule to the rule set.

addRule(SingletonRuleSet ruleset) *SingletonRuleSet* objects may contain additional *SingletonRuleSet* objects nested as deeply as you care to go. This provides a way to merge the results of several rule sets.

evaluate(double value) All of the rules in this set are evaluated against *value* and the results are merged together. The result of any nested rule sets are also combined with this accumulated result.

Note that nested rule sets must have had their *evaluate()* method called first, to generate a result. Nested sets are not evaluated against this value because they may, in fact, operate on a different input.

evaluate( ) All of the rule sets, but not the rules, have their results combined into a result. These rule sets should have had their *evaluate()* called with the appropriate input value prior to this call.

defuzzy( ) The accumulated result associated with this rule set is defuzzified according to the singleton formula introduced in Equation 3-7.

```
for (int idx=0; idx<num; idx++)
{
   value += m_result.getX(idx) * m_result.getY(idx);
   weight += m_result.getY(idx);
}
if (MathConst.isZero(weight))
{ return 0.0; }

return value/weight;
```

Code: Fuzzy Heater Controller

Now that the basic fuzzy system has been introduced and defined we can use it to create a heater controller that roughly mirrors the PID controller.

This controller is going to use two rule sets, one for the proportional part of the control and the second for the integral part of the control.

If we did this according to the grid in Figure 3-7 we would define 25 rules, each with two conditions and one result.

However, our rules do not provide for multiple conditions. How can we do this? First, we will not have to for reasons we explore in a minute. But if we did have to, we could combine the results of two rules using fuzzy operators. These operators are explored in more detail later.

For example, the rule "if (e is warm) and (ei is negsmall) then output is medium" could code as:

```
FuzzySet w = new TrapezoidSet( 35, 10, 70 );
SingletonRule is_w = new SingletonRule( warm, 10 );

FuzzySet ns = new TriangleSet( -30, 60 );
SingletonRule is_ns = new SingletonRule( negsmall, -50 );

FuzzySet w_and_ns = FuzzyOp.and(is_w.evaluate(e),
                    is_ns.evaluate(ei));
control = w_and_ns.getCOG();
```

Or we could invent a *Rule* interface that takes multiple parameters.

However, we can avoid this rule explosion using something known as *Comb's Method*, named after William E. Combs. In his 1997 paper *The Combs Method for Rapid Inference*, William Combs showed that fuzzy rules like the one in Equation 3-10 could be transformed into the form shown in Equation 3-11.

$$[(p \text{ and } q) \text{ then } r] \qquad \textbf{3-10}$$

$$[(p \text{ then } r) \text{ or } (q \text{ then } r)] \qquad \textbf{3-11}$$

This means we do not have to build the 25 rules in a *Rule* interface that allows multiple inputs. Instead, we can build 10 rules (five for each input) and combine their results using our current *Rule* class. This is shown in more detail below.

Fuzzy.java

aip.app.fuzzy.Fuzzy The *Fuzzy* test program has a lot of bulk devoted to the user interface. We also introduce the *JLineGraph* class here, which is a rescaling graph that basically draws lines between datum points. Its class hierarchy is shown in Figure 3-17.

The *Fuzzy* test creates three displays. The top two show the shapes of the membership sets and the bottom one is a scrolling chart that shows the set-point and the heater response as driven by the controller. This is shown in Figure 3-18. Other than the extra graphs and the inclusion of fuzzy rules, the *Fuzzy* controller is the same as the *PID* controller.

A summary of *Fuzzy* is given in Table 3-8.

The control rules in this test make no effort to exactly mimic the control curve of the PI controller. They were simply adjusted to provide "good enough" behavior. According to theory, the fuzzy controller can be made to exactly mimic the PID controller.

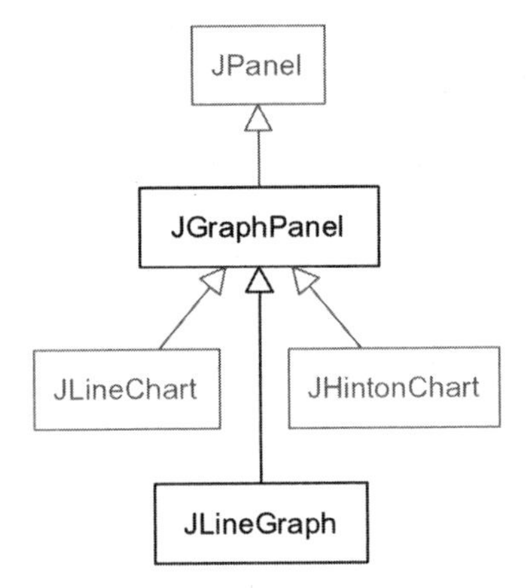

Figure 3-17 JLineGraph class diagram

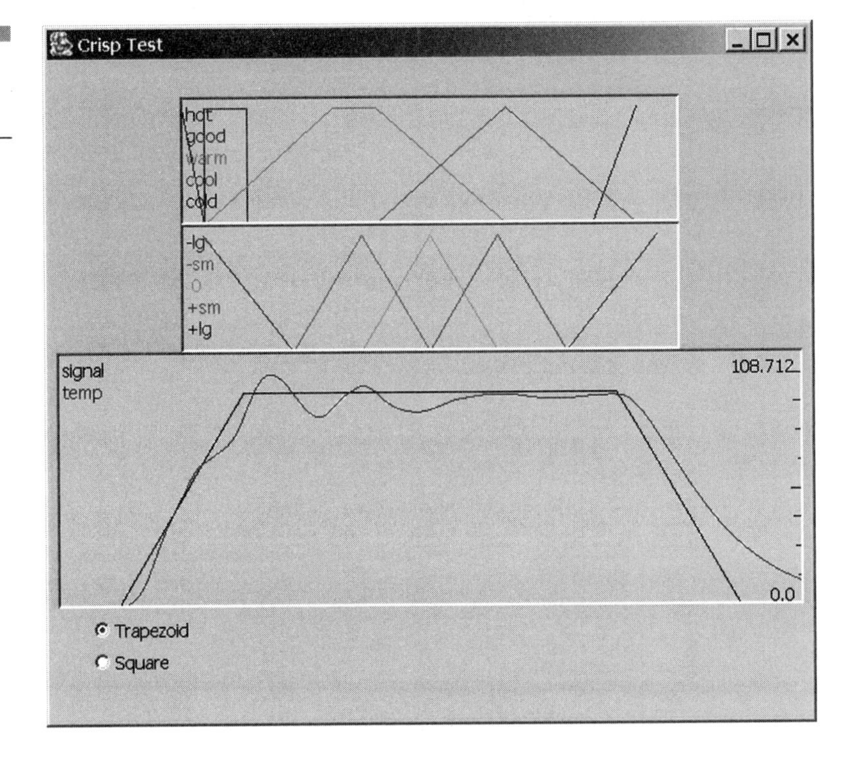

Figure 3-18 Fuzzy controller display

assemble() This goes through the laborious process of building the UI.

run() The *run()* method starts by creating all of the fuzzy rules and rule sets used by the controller. There are two sets of singleton rules in this

Table 3-8 Fuzzy controller summary

Main Loop	
void	assemble ()
void	run ()
Process signal generators	
double	signal_square (double t)
double	signal_trapezoid(double t)
Fuzzy Controller	
double	do_fuzzy_p (double err)
double	do_fuzzy_pi (double err)

controller. The first generates a control signal based on the error in temperatures, and follows the form:

if *e* is hot then output 0
if *e* is good then output 5
if *e* is warm then output 10
if *e* is cool then output 20
if *e* is cold then output 100

These rules are created and drawn using the following code:

```
FuzzySet hot = new RampSet(0, -5);
FuzzySet good = new SquareSet( 5, 10 );
FuzzySet warm = new TrapezoidSet( 35, 10, 70 );
FuzzySet cool = new TriangleSet( 70, 60 );
FuzzySet cold = new RampSet( 90, 100 );

hot.draw(m_graph1, 0);
good.draw(m_graph1, 1);
warm.draw(m_graph1, 2);
cool.draw(m_graph1, 3);
cold.draw(m_graph1, 4);

SingletonRule is_hot = new SingletonRule( hot, 0 );
SingletonRule is_good = new SingletonRule( good, 5 );
SingletonRule is_warm = new SingletonRule( warm, 10 );
SingletonRule is_cool = new SingletonRule( cool, 20 );
SingletonRule is_cold = new SingletonRule( cold, 100 );

m_rate = new SingletonRuleSet();
m_rate.addRule( is_hot );
m_rate.addRule( is_good );
m_rate.addRule( is_warm );
m_rate.addRule( is_cool );
m_rate.addRule( is_cold );
```

Note that the rules in code are much bulkier than the English prose they implement. A usable fuzzy library could include standard definitions for many types of sets as well as a parser that converts conversational rules into formal code. This is not one of those libraries.

Likewise, the integrated error rules are defined in code as:

```
FuzzySet neglarge = FuzzyOp.scale(new RampSet( -60, -100 ),
                    IFACTOR);
FuzzySet negsmall = FuzzyOp.scale(new TriangleSet( -30, 60 ),
                    IFACTOR);
FuzzySet zero = FuzzyOp.scale(new TriangleSet( 0, 60 ), IFACTOR);
FuzzySet possmall = FuzzyOp.scale(new TriangleSet( 30, 60 ),
                    IFACTOR);
FuzzySet poslarge = FuzzyOp.scale(new RampSet( 60, 100 ), IFACTOR);

neglarge.draw(m_graph2, 0);
negsmall.draw(m_graph2, 1);
zero.draw(m_graph2, 2);
possmall.draw(m_graph2, 3);
poslarge.draw(m_graph2, 4);

SingletonRule is_neglarge = new SingletonRule( neglarge, -200 );
SingletonRule is_negsmall = new SingletonRule( negsmall, -50 );
SingletonRule is_zero = new SingletonRule( zero, 0.0 );
SingletonRule is_possmall = new SingletonRule( possmall, +40 );
SingletonRule is_poslarge = new SingletonRule( poslarge, +100 );

m_accel = new SingletonRuleSet();
m_accel.addRule( is_neglarge );
m_accel.addRule( is_negsmall );
m_accel.addRule( is_zero );
m_accel.addRule( is_possmall );
m_accel.addRule( is_poslarge );
```

The new statement *FuzzyOp.scale(FuzzySet set, double factor)* provides a scaled version of a set, changing its maximum membership weight by *factor* as documented in the next section. What this does is reduce the importance of the integrated error in the final output.

Once the rules are built they are assembled into a rule set.

```
m_power = new SingletonRuleSet();
m_power.addRule( m_rate );
m_power.addRule( m_accel );
```

Note that this program can be compiled to run either *do_fuzzy_p()* or *do_fuzzy_pi()* control methods.

signal_square(double t)
signal_trapezoid(double t) These methods generate a set-point signal that the controller tries to match.

do_fuzzy_p(double err) Perform the fuzzy control using only the proportional error term and the *m_rate* rule set.

```
m_rate.evaluate(m_e);
m_out = m_rate.defuzzy();
```

do_fuzzy_pi(double err) Perform the fuzzy control using both sets of rules.

```
m_ei *= ILEAK;
m_ei += m_e;

m_rate.evaluate(m_e);
m_accel.evaluate(m_ei);

m_power.evaluate();

m_out = m_power.defuzzy();
```

The integrated error was implemented with a leak. This reduces wind-up and makes the controller more stable.

Code: Fuzzy Logic

This final section describes the rest of the fuzzy logic library provided with this book. The final class diagram is shown in Figure 3-19. It adds a utility class *FuzzyOp* that modifies *FuzzySet* objects, plus the *FuzzyRule* and *FuzzyRuleSet* classes that generate a shape for their result instead of a singleton.

FuzzyRule.java

aip.fuzzy.FuzzyRule A *FuzzyRule* is the same as a *SingletonRule* with the exception of its output. It creates a scaled fuzzy set as its result rather than the specialized singleton set.

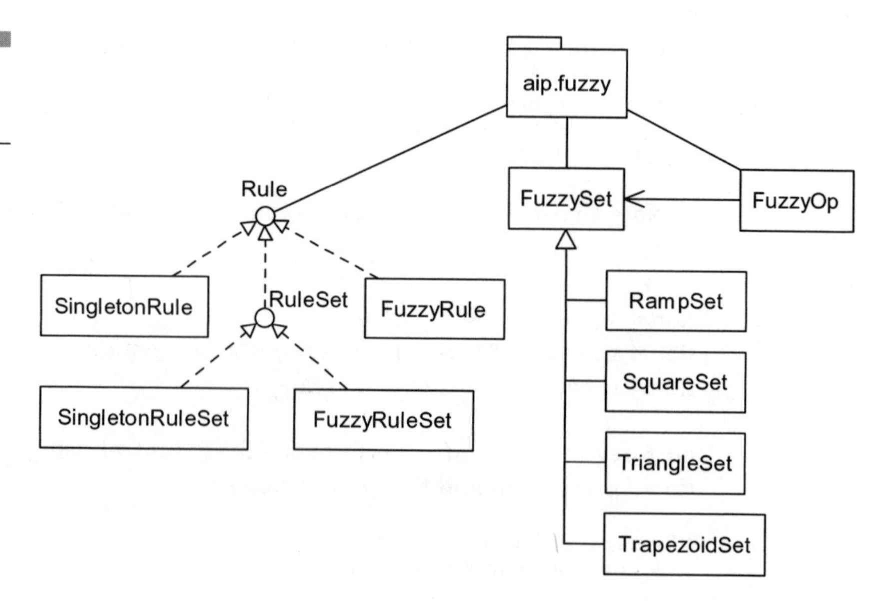

Figure 3-19 Fuzzy logic class diagram

A summary of *FuzzyRule* is given in Table 3-9.

FuzzyRule(FuzzySet condition, FuzzySet result) Generate a *FuzzyRule* with the specified condition and result.

evaluate(double value) Evaluate the rule against an input, generating a *FuzzySet* result that is scaled by the membership weight of *value*. Note that another way of weighting the result set is through truncation instead of scaling. Figure 3-20 shows the difference. Scaling the result set reduces its area more, so it loses its influence faster.

If the value does not fall in the range of the membership set, *evaluate()* returns *null*.

result() Return the result calculated by the previous call to *evaluate()*.

defuzzy() Generate a crisp value by calculating the center of gravity of the result set. Note that *FuzzySet* performs the actual COG calculation.

FuzzyRuleSet.java

aip.fuzzy.FuzzyRuleSet A *FuzzyRuleSet* is a collection of *FuzzyRule* and *FuzzyRuleSet* objects. While you could technically add the singleton rules it is not recommended.

A summary of *FuzzyRuleSet* is given in Table 3-10.

FuzzyRuleSet() Construct an empty rule set.

addRule(Rule rule) Add a rule to the rule set.

Table 3-9
FuzzyRule summary

Public Methods	
	FuzzyRule (FuzzySet condition, FuzzySet result)
FuzzySet	evaluate (double value)
FuzzySet	result ()
double	defuzzy ()

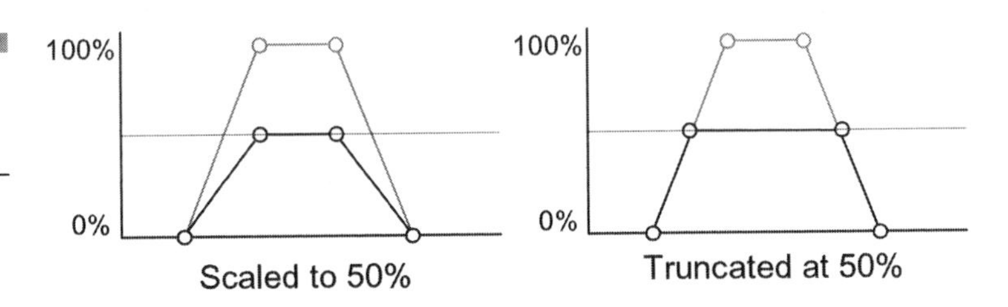

Figure 3-20
Scaling versus truncation

Table 3-10

FuzzyRuleSet summary

```
Construction
          FuzzyRuleSet ()
    void  addRule (Rule rule)
    void  addRule (RuleSet ruleset)

Evaluation
FuzzySet  evaluate (double value)
FuzzySet  evaluate ()
FuzzySet  result ()
  double  defuzzy ()
```

addRule(RuleSet ruleset) *FuzzyRuleSet* objects may contain additional *FuzzyRuleSet* objects, nested as deeply as you care to go. This provides a way to merge the results of several rule sets.

evaluate(double value) All of the rules in this set are evaluated against the input *value* and the results are merged together using *FuzzyOp.or()*. The results of any rule sets are also merged into this accumulated result.

The *or* process is not efficient, since it allocates a new *FuzzySet* with each operation. A more efficient set of fuzzy operators would also have more complicated code. For this book I have erred on the side of understandability, sacrificing efficiency where needed.

Note that nested rule sets must have had their *evaluate()* method called first, to generate a result. These sets are not evaluated against this value because they may, in fact, operate against a different input.

evaluate( ) All of the rule sets, but not the rules, have their results combined into a final result using *FuzzyOp.or()*. These nested rule sets should have had their *evaluate()* called with the appropriate input value prior to this call.

defuzzy( ) The accumulated result associated with this rule set is defuzzified using the *FuzzySet.getCOG()* method.

There are other ways that a set can be defuzzified other than center of gravity, and many of these methods are explored in the books in the *Additional Reading* section.

FuzzyOp.java

aip.fuzzy.FuzzyOp *FuzzyOp* is a utility class of static methods that manipulate fuzzy sets. Some of the methods modify a single set to change its shape and other methods take two sets and combine them in some way. All of these methods create a new *FuzzySet* as the result of the operation.

A summary of *FuzzyOp* is given in Table 3-11.

not(FuzzySet in) Create a new fuzzy set that is the inverse of the specified set (Figure 3-21).

scale(FuzzySet in, double scale) Create a new fuzzy set with membership weights scaled by *scale* (Figure 3-22).

Table 3-11
FuzzyOp summary

Modifiers and Hedges

```
FuzzySet not (FuzzySet in)
FuzzySet scale (FuzzySet in, double scale)
FuzzySet normalize (FuzzySet in)
FuzzySet barely (FuzzySet in)
FuzzySet somewhat (FuzzySet in)
FuzzySet very (FuzzySet in)
FuzzySet extremely (FuzzySet in)
```

Binary Operators

```
FuzzySet and (FuzzySet alpha, FuzzySet beta)
FuzzySet or (FuzzySet alpha, FuzzySet beta)
FuzzySet add (FuzzySet alpha, FuzzySet beta)
FuzzySet mult (FuzzySet alpha, FuzzySet beta)
```

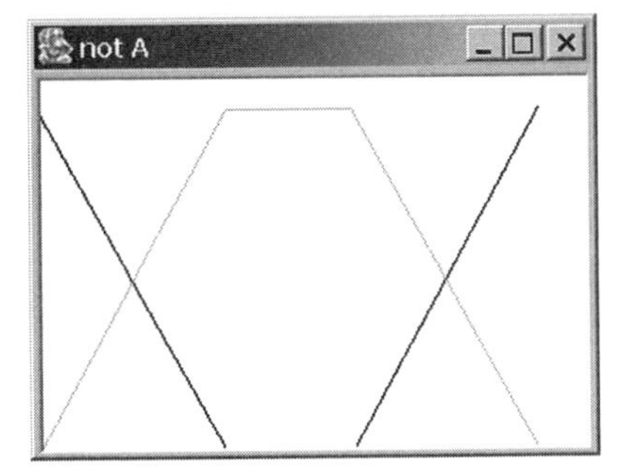

Figure 3-21
not A

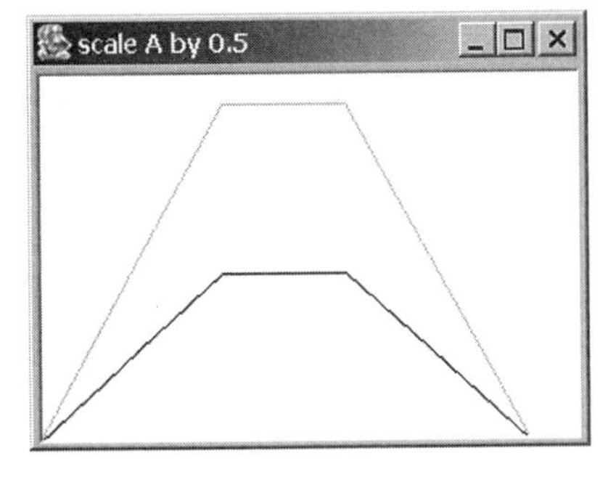

Figure 3-22
scale A by 0.5

normalize(FuzzySet in) This is an inverse of the *scale()* operator. This method scales the set to the largest membership value of 1.0, scaling the set up or down as needed.

barely(FuzzySet in) Create a new fuzzy set that is a bloated version of the specified set (Figure 3-23). Even when a value is barely in the set, it has a large membership weight.

This modification of a fuzzy set is called hedging, and provides a way to tune a membership function to better match your intent.

somewhat(FuzzySet in) Create a new fuzzy set that is a somewhat more inclusive version of the specified set (Figure 3-24).

extremely(FuzzySet in) Create a new fuzzy set that is a constricted version of the specified set (Figure 3-25). A value has to be extremely near the heart of the set before the membership value gets large.

very(FuzzySet in) Create a new fuzzy set that is a somewhat less inclusive version of the specified set (Figure 3-26).

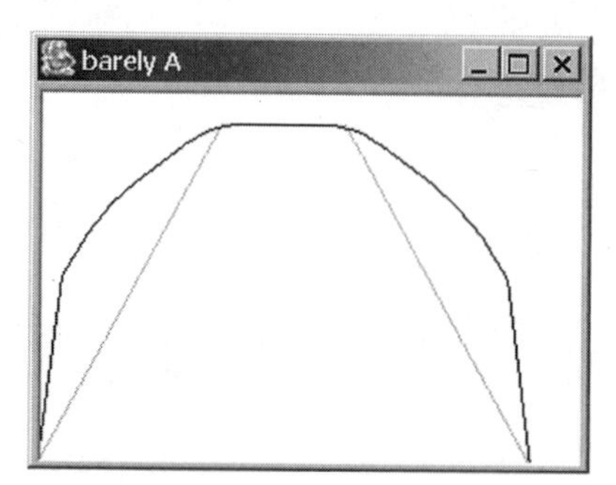

Figure 3-23 barely A

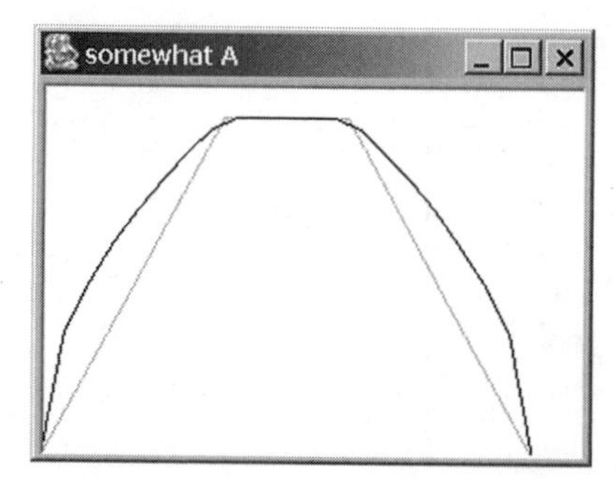

Figure 3-24 somewhat A

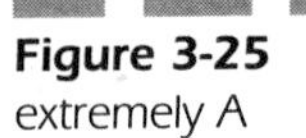

Figure 3-25
extremely A

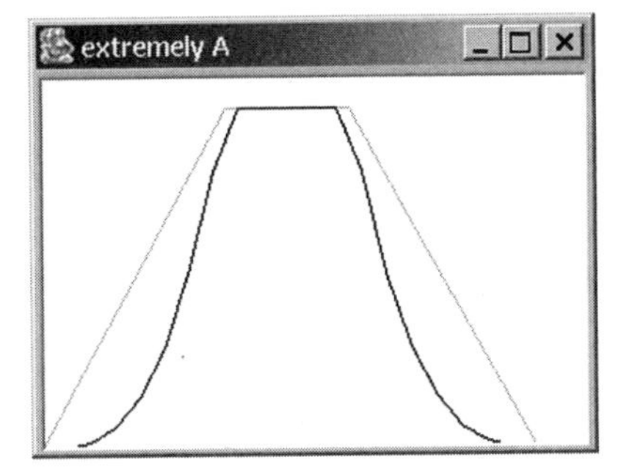

Figure 3-26
very A

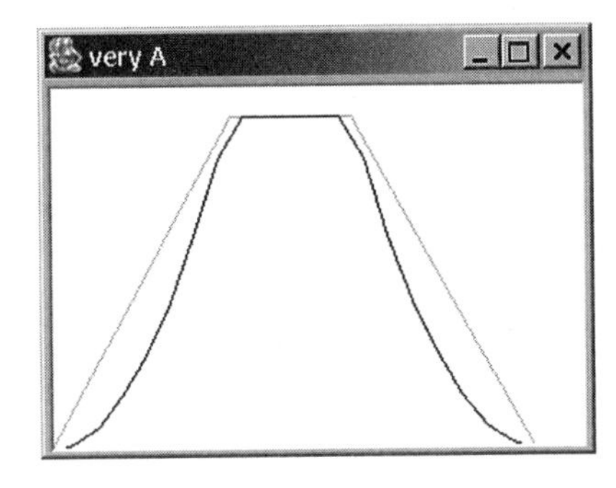

Figure 3-27
A and B

and(FuzzySet alpha, FuzzySet beta) Create a new fuzzy set that is the intersection of the two specified sets (Figure 3-27).

or(FuzzySet alpha, FuzzySet beta) Create a fuzzy set that is the union of the two specified sets (Figure 3-28).

add(FuzzySet alpha, FuzzySet beta) Create a fuzzy set that is the sum of the two specified sets (Figure 3-29).

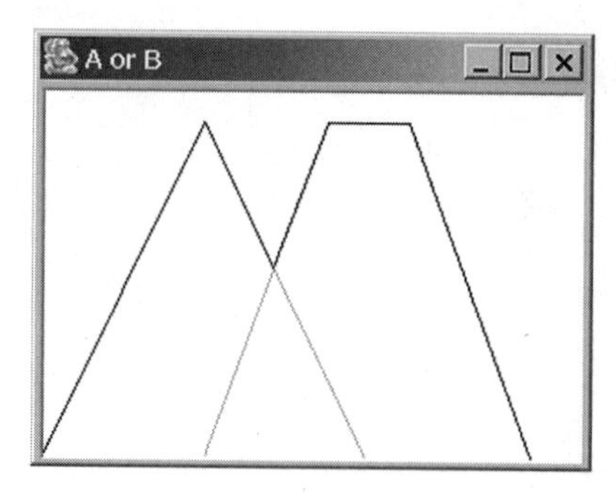

Figure 3-28
A or B

Figure 3-29
A add B

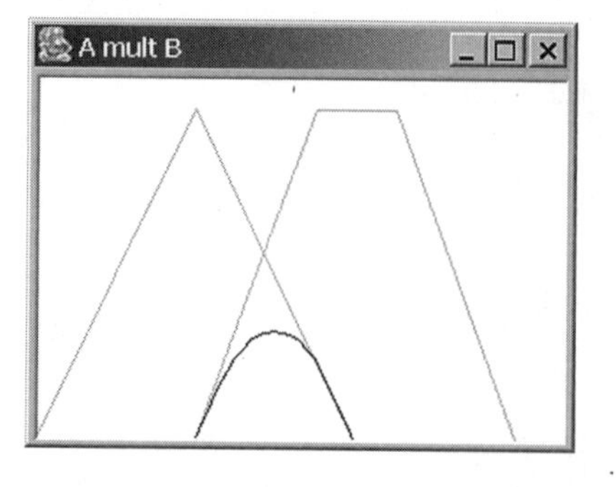

Figure 3-30
A mult B

mult(FuzzySet alpha, FuzzySet beta) Create a fuzzy set that is the factor of the two specified sets (Figure 3-30).

Additional Reading There are many resources available on fuzzy logic and fuzzy systems, plus a number of free fuzzy libraries and code examples.

The IEC has a standard for industrial fuzzy controllers, an older version of which can be found freely on the internet (IEC 1131, 1997).

There is an open source fuzzy software project on sourceforge, at ffll.sourceforge.net.

For a good introduction to the ideas behind fuzzy logic, *Fuzzy Thinking* by Bart Kosko is an excellent book. For advanced concepts in fuzzy logic, *Fuzzy Engineering* is the book to find (Kosko, 1997).

There are many practical books available, such as *The Fuzzy Systems Handbook* (Cox, 1994) and *Fuzzy Logic & NeuroFuzzy Applications Explained* (Altrock, 1995), plus many others.

Neural Reflexes

Here is a quick peek at some other possibilities for reflexive control. We will explore neural networks in detail later.

Computational neurons, as typically used by neural systems, are simple software or hardware devices that take any number of input signals, scale these signals by their relative importance to the neuron (the signal weight factor), sum the signal, and test it against a threshold or transfer function. The output of the neuron is then a factor of the sum of the weighted inputs, as shown in Figure 3-31. If this is a threshold neuron, the output will be all on or all off. Transfer functions ramp the input from zero to one along a smooth curve.

When a weight is positive, the input acts as a stimulus to the neuron. However, if you allow weights to be negative, this creates an inhibitory input to the neuron.

A simple neuron with two inputs, one excitatory and one inhibitory, could be created with a single operational amplifier and a handful of resistors to implement the weights. The threshold would be adjusted with a bias on the negative input. If this is meaningless to you, do not worry about it.

Even simple hard-wired neurons can be used to create behavior, as demonstrated by Valentine Braitenberg in the mid 1960s (Braitenberg, 1984).

Figure 3-31 Computational neuron

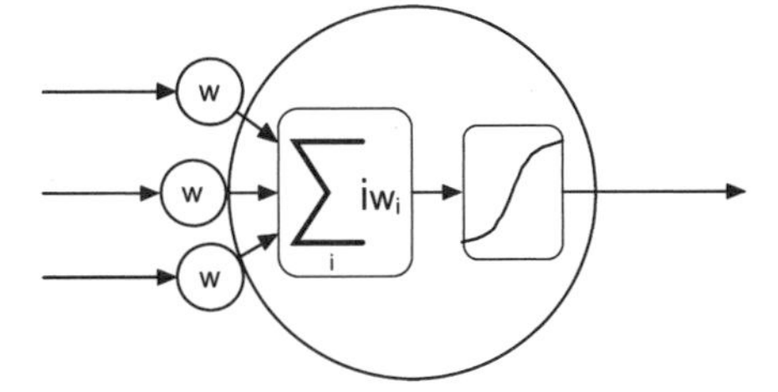

An interesting tool to explore behavior and neural networks is the BugBrain game, which you can find on the website www.biologic.com.au/bugbrain, as well as on www.simtel.net and www.winsite.com.

Autoassciative neural nets can be used for filtering, too (Masters, 1993). These networks learn how the signal is supposed to look and then, given a noisy version of that signal, they can reproduce the clean version.

Not all systems require fancy control systems and many fancy control technologies can be adapted down to simple reflexes.

CHAPTER 4

Scripted Behavior

This chapter looks at the subject of data-driven intelligence. Or, more accurately, ways a program can pretend to be intelligent, while actually being as dumb as a post, by reading off of really clever crib sheets.

First we look at what we mean by data-driven intelligence. Then finite state machines are explored in detail. These machines provide a framework for a wide variety of stimulus–response systems such as appliances, games, and other practical applications.

Taking the basic state machine we start rolling dice with the stochastic Markov model. Finally we take a brief tour through the world of chatbots and frame-based dialogs.

Data-Driven Intelligence

First, what do I mean by "data driven"? When a system is data driven it consists of a somewhat generic "machine" whose behavior is determined by some external template. The external template is the data.

Compilers are not necessarily data driven. The source code that they process is *input* and not a template of behavior.

However, there are assemblers and compilers available that are data driven. These use a template that describes the machine they compile to. The compiler plus this template can process source code and create a correct result. If you wanted to make a compiler for a new computer you write a new template that describes that computer's architecture instead of re-writing the whole compiler.

This is the benefit of a data-driven system. When you want it to have different behavior you update the data that defines its behavior instead of going into the source code and re-writing.

While many different programs can be data driven this chapter looks at state-based systems. Input arrives in the form of some event, user input, or other action. This input is tested against current possibilities and some response is chosen.

There are two friendly ways the data can be created. One way is to use text files that are parsed into the appropriate internal structures. These could be parsed all at once at startup or read incrementally as needed.

The other way is to create a tool that edits machine-readable data files, moving the difficulties of parsing outside the run environment. These binary files are also more secure, making it harder for the end user to manipulate the system. This could be either good or bad, depending on your goals.

The examples in this chapter use neither of these methods. To keep the code to a minimum we skip all forms of parsing and create the data structures directly in code. This removes much of the benefit of being data driven, but it also removes the need to develop and explain a parsing front-end.

Finite State Machines

Finite state machines (FSMs), also known as finite state automata or transition networks, are very simple. The elements of an FSM are the state, the arc or transition, input events, and actions. An FSM diagram showing these elements is shown in Figure 4-1.

Starting with some initial state (State 1, in this case), the machine receives events. Each arc leaving a state has a condition associated with it. When an event matches the arc's condition the machine is shifted along that arc to its destination state. If no condition matches nothing happens.

If there is an action associated with the arc it will execute when traversing the arc. There may be actions associated with states, or the states may simply be states. The machine terminates if it transitions to a terminal state.

States have names or some type of identifier. If a state contains an action a common notation is *Name:Action*, separating the name and action with a colon. The same goes for arcs. An arc may be labeled *Condition* or *Condition:* for an actionless arc, and *Condition:Action* when an action is invoked. An arc may trigger no matter what the event is, in which case it may specify just an action:*Action*.

State machines can be used for a wide range of applications. They are well suited to event-driven systems such as a vending machine coin acceptor, a calculator, or computer-game AI.

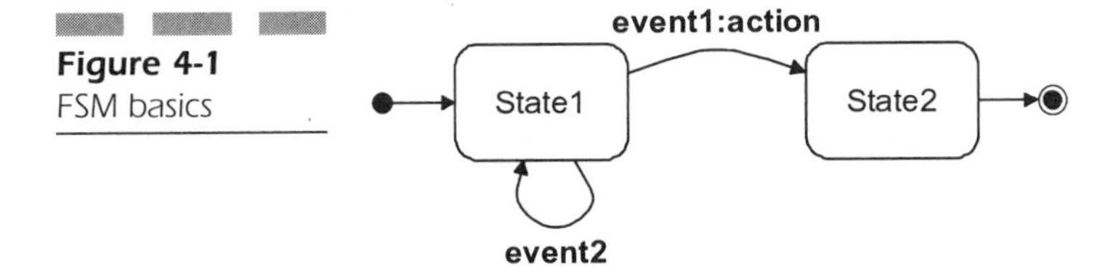

Figure 4-1
FSM basics

FSMs are used as a hardware design tool in one of two forms, the Moore machine or the Mealy machine. The distinctions do not concern us here, but make a difference when you are converting your FSM diagram into electronics.

FSMs are equivalent in power to regular grammars, though not necessarily as powerful as context-free grammars (CFG). The addition of recursion, making the FSM a recursive transition network, or RTN, kicks FSMs up to CFG strength (Allen, 1995).

State machines come in different forms. Acceptors are used to recognize or accept patterns of input. Recognizers are similar to acceptors but instead of a simple yes/no answer they categorize the inputs. Transducers use the input to generate some output. In a similar vein, state machines can be used to control an external process.

Designing an FSM

Though Katz (1993) is focused on the hardware implementation of finite state machines, he provides a good introduction to the subject of FSM design as well as a coin acceptor example.

The first step in designing an FSM is the same as the first step in any project. You need to understand the problem you are trying to solve and have some idea of how you want to solve it. Write out the problem on a paper and explore various ways of approaching it. For our example we want to make a machine that accepts nickel and dimes. When 15 cents or more has been put into the machine it opens a door. Maybe this is a vendor for cheap newspapers.

Once you have an idea of the problem you can create an FSM diagram showing the possibilities. Given our coin box problem this preliminary FSM might look like Figure 4-2. In this diagram we take a different arc for each nickel *N* or dime *D* inserted in the machine, until we reach a state where the box is open.

The third step is to take your initial diagram and simplify it. In Figure 4-2 there are extra states that can be merged, especially the five *Open* states. It can help to think of each state as the amount of money received so far, giving the reduced diagram shown in Figure 4-3.

If we wanted to associate the actions with the transition arcs instead of with the state, it would look like Figure 4-4.

The fourth and final step is to implement the machine in software or hardware. We will look at several ways to implement this state machine.

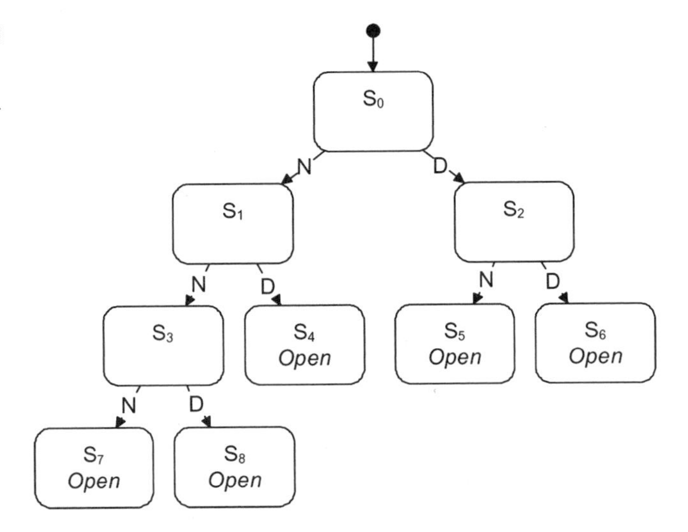

Figure 4-2 Coin box FSM

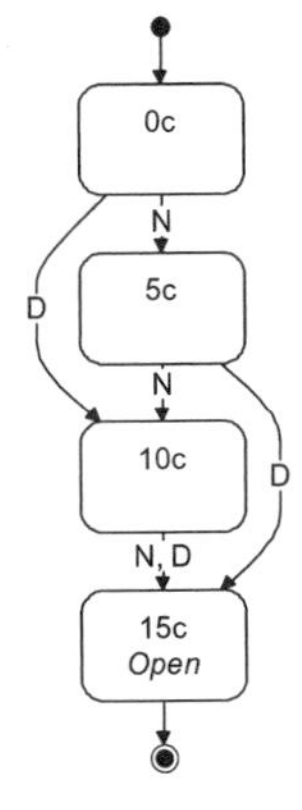

Figure 4-3 Simplified coin box FSM

One way that avoids the data driven aspect entirely is to code the states and transitions in a nested if/then chain. Something along the lines of:

```
if (state == 0)
{
   if (event == 0) then { ... }
   else if (event == 1) then { ... }
   ...
}
else
{
...
}
```

But we do not do that here.

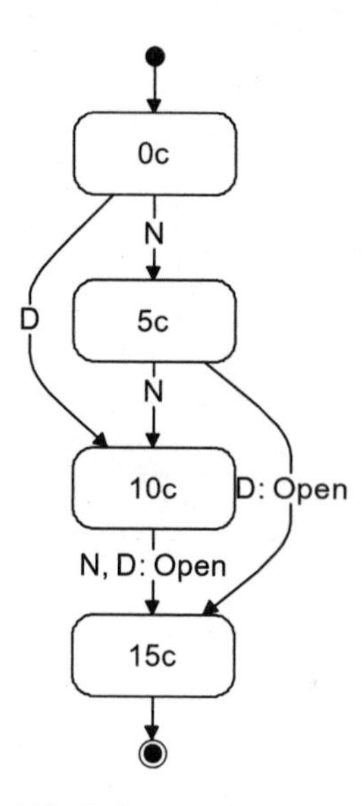

Figure 4-4 Alternate coin box FSM

Table Driven FSM

In the simplest case a state machine consists of a simple lookup method and a two-dimensional table of transitions. Each state would be a subset of the table. For each sub-table there is a transition for each possible event. Such a table might look like Table 4-1.

The *Current State* and *Event* columns could both be implied based on their position within the table. The *Action* and *Next State*, however, must be specified. And even the *Action* could be left off if there is an external observer performing the actions based on the machine's state.

To be really data driven, the table should exist as an external file that is read into the application.

In some environments you can overload your compiler's pre-processor so that it parses a state language for you. The statements in the FSM language would be defined as pre-processor macros and generate if/then chains in source code. An example of this can be seen in Rabin (2000).

Back to the table. Assuming the state, action, and event are all defined as *byte* values, the table might be coded in Java something like this:

```
Arc[][] table = new Arc[][]
{
    /* State 0: 0c */
    new Arc[]
    {
        new Arc(1, 0), /* Nickel */
        new Arc(2, 0), /* Dime */
    },
```

Table 4-1

Coin box FSM table

Current state	Event	Action	Next state
State 0c	Nickel Dime	n/a n/a	5c 10c
State 5c	Nickel Dime	n/a Open	10c 15c
State 10c	Nickel Dime	Open Open	15c 15c
State 15c	n/a	n/a	n/a

```
        /* State 1: 5c */
        new Arc[]
        {
            new Arc(2, 0), /* Nickel */
            new Arc(3, 1), /* Dime */
        },
        /* State 2: 10c */
        new Arc[]
        {
            new Arc(3, 1), /* Nickel */
            new Arc(3, 1), /* Dime */
        },
        /* State 3: done */
        null
    };
```

Code to process events against this table might look something like this:

```
class Arc
{
    public byte dest;
    public byte action;
}
class FSM
{
    public FSM( Arc[][] table )
    {
        m_table = table;
        m_state = 0;
    }
    public void event(byte event)
    {
        byte action = m_table[m_state][event].action;
        do_action(action);
        m_state = m_table[m_state][event].dest;
    }
    private byte m_state;
    private Arc m_table[][];
}
```

This code is presented as an example. The next section builds a more complex version of the table-driven FSM.

Code: Table-Driven FSM

This table-driven FSM application implements a variation of the table described in the previous section. It implements the machine shown in Figure 4-5. This machine is somewhat expanded from Figure 4-4.

The table is not created with undifferentiated bytes, but uses *state* and *transition* classes to manage the data. A *state* is a named collection of *transitions* and a *transition* holds an event character, a destination state, and an optional action.

The events are represented as characters and include "d" for dropping a dime into the machine, "n" for a nickel, and "t" to take the product from the box.

Until product is taken, the machine will continue eat money.

Actions are represented as strings and are printed to the console.

FSMTable.java *FSMTable* is a command line program with no user interface worth mentioning. It is invoked with one or more parameters that are character sequences to be processed. For example:

```
java aip.app.fsmtable.FSMTable nnnt nndt dddt
```

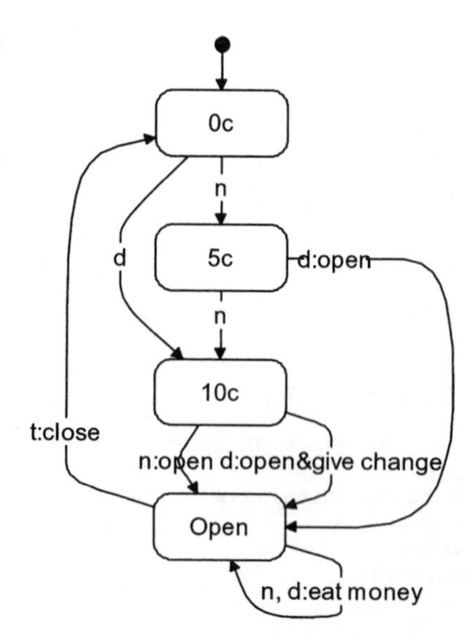

Figure 4-5 Table driven coin box diagram

This will run three sequences through the machine. The first sequence is "nickel, nickel, nickel, take product." You can probably figure out the other two.

The result of this run is shown here:

```
-----------------------------
Process: 'nnnt'
state: 0c
(n) state: 5c
(n) state: 10c
(n) - Open
state: OPEN
(t) - Close
state: 0c
-----------------------------
Process: 'nndt'
state: 0c
(n) state: 5c
(n) state: 10c
(d) - Open and Give Change
state: OPEN
(t) - Close
state: 0c
-----------------------------
Process: 'dddt'
state: 0c
(d) state: 10c
(d) - Open and Give Change
state: OPEN
(d) - (eat money)
state: OPEN
(t) - Close
state: 0c
```

aip.app.fsmtable.FSMTable The *FSMTable* application is a self-contained command-line application that processes nickels and dimes in our hypothetical coin-accepting machine.

A summary of *FSMTable* is given in Table 4-2.

Table 4-2

FSMTable summary

Public Methods

```
     FSMTable ()
void run(String input)
```

Static Public Methods

```
void main(String args[])
```

Private Methods

```
 int event(char token)
void setState(int state)
```

main(String args[]) The *main* method actually has a small job in this application. It first creates an *FSMTable* object and then it iterates through its arguments, passing them one at a time to the *run()* method.

FSMTable( ) The constructor does the work of instantiating the state and transition data. It is not very pretty, but it gets the job done.

run(String input) The *run()* method is the outside loop of our little machine. It starts up with some initialization and then dives into the *input* string. It feeds this string, forced to lower-case, one character at a time to the *event()* method. The result of *event()* is a new state, so this is sent to *setState()*. The code that does this demonstrates the simplicity of state machines.

```
int num = input.length();
for (int idx=0; idx<num; idx++)
{
    setState(event( Character.toLowerCase(input.charAt(idx)) ));
}
```

event(char token) *event()* takes a single event code, in this case a lower-case character, and determines which state transition it matches. It then returns the destination state from that transition, printing any action string along the way.

Each transition is checked in turn to find the first match. There is an implied priority to the transitions, based on their order in the list. This priority ordering is present in most state machine implementations.

The event code is simple:

```
int num = m_state.arc.length;
for (int idx=0; idx<num; idx++)
{
    if (m_state.arc[idx].event == token)
    {
        String action = m_state.arc[idx].action;
        if (action != null)
        {
            System.out.println(" - "+action);
            System.out.print(" ");
        }
        return m_state.arc[idx].state;
    }
}
```

setState(int state) This method remembers the given state index. It also prints out the state's name to the console so that you can follow along.

Code: FSM Library

We make a number of enhancements to the basic table-driven state machine at the expense of added complexity. Even though the state machine described here is complex, it is not all that complex. The basic operation is the same as above, just with more frills and more flexibility. Here is a list of some enhancements that are or can be made to the state machine processor:

- *Multiple machines.* Only the simplest environments will need only a single state machine generating its behavior. By placing an event manager at the head of the FSM system we can send messages to any number of different state networks.

- *Context for communication.* Actions do not usually exist in a vacuum. They need to affect their environment and this environment is their *Context*. The context can be modified by actions and tested by event conditions, as needed. The *Event* itself is part of the context so it may also be used by actions and conditions. Note that the *SymbolTable* class is an interpreted variable list. A context could be created with a symbol table to manage arbitrary variables.

- *Delayed actions*. Not all events occur immediately. A delayed event is like a note for the future. "At such a time, self destruct", or "after 10 seconds, timeout and give up". Delayed events add a new dimension to the system.

- *Application specific events, conditions, and actions*. Each application has unique needs, so a generic FSM library needs to provide a way to specify what an event is, the conditions to test in order to traverse an arc, and what actions perform.

- *Communication between machines.* Actions in one machine may need to send events to a different machine, or send events or delayed events to itself.

- *Actions on states and transitions*. Some machines perform actions on the arc transitions; other machines perform actions upon entering a state. Some do both.

- *Multiple actions.* Sometimes you need to perform multiple actions on a state or transition, not just one action.

- *Recursive networks*. Though not explicitly explored here, it should be possible to manipulate the machine so that one network can call another network. This way an entire network can stand in for a single state in another network, or one network can recurse.

- *Series of machines*. When one state machine finishes its operation it could change the default machine and send it a start-up message. Behaviors can be chained together to make even more complex behaviors.

- *Error handling*. Though not explored here, errors can be managed through the context or as special actions or states in the network.

- *Default states*. Though there is a default machine that un-targeted events get sent to, there is no default state for a given state machine. It is possible to create a "fall-through" state in a machine. If the current state does not handle an event that event would be sent to this fall-through state. In case of a match that state would be entered. This is another way to handle error or termination states.

- *Event scoping.* Events may be targeted at a specific state machine or at a given state so that any other state will ignore that event. This can be useful for a delayed *timeout*: event that is only valid for the state that sent it.

The class diagram for this new, expanded FSM library is shown in Figure 4-6.

Note that in this code, like in all the code for this book, error testing has been kept to a minimum. Production code should be much more paranoid.

There are a lot of classes here, not to mention the *Context*, *Condition*, and *Action* interfaces that need to be implemented before anything else can be done. The gist of the system, however, is simple.

The *EventQueue* is the root of it all and contains any number of *Machine* and *Event* objects.

A *Machine* consists of one or more *State* objects.

A state may contain an *Action* and will have one or more *Transition* objects.

Finally, a *Transition* consists of a *Condition* and an *Action*.

EventQueue.java

aip.fsm.EventQueue The *EventQueue* manages one or more FSMs and handles the events for all of these machines.

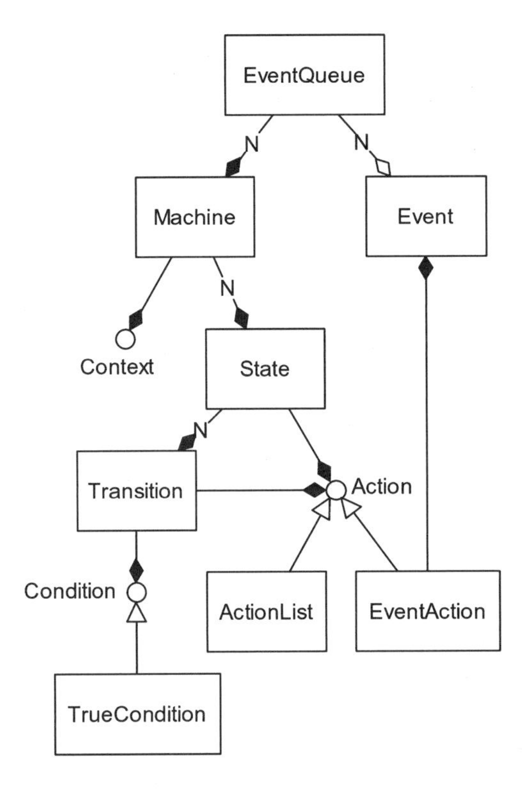

Figure 4-6 FSM library class diagram

Table 4-3 EventQueue summary

```
Construction
        EventQueue    getInstance ()

Machine Management
    int  countMachine()
Machine  newMachine(String name)
Machine  getMachine(String name)
Machine  removeMachine (String name)
Machine  getDefault()
Machine  setDefault(String name)

Event Processing
   void  tick()
boolean  sendNow(Event event)
   void  sendDelay(Event event, long delay)
```

A summary of *EventQueue* is given in Table 4-3.

getInstance() The *EventQueue* is written as a Singleton, which means that there will only ever be one instance of this class. The static *getInstance()* method is used to create and then retrieve this instance.

countMachine( )
newMachine(String name)
getMachine(String name)
removeMachine(String name) The *EventQueue* is the portal to the state machines. These machines are created, queried, and manipulated by these methods.

getDefault( )
setDefault(String name) There may be a default *Machine* in the *EventQueue*. This machine receives any events that do not specify an explicit recipient.

tick( ) Some events can be sent with a time delay.

The client code controls the update interval. The client needs to call *tick()* at some regular interval in order for queued events to get sent at their time. Of course, if all events are sent using *sendNow()*, *tick()* never needs to be called.

sendNow(Event event) Send an event to a machine without any delay. The event gets sent to its targeted machine if it specifies its destination, otherwise it will get sent to the default machine.

sendDelay(Event event, long delay) Send an event to a machine with the specified delay. The delay is in milliseconds, which are 1/1000 of a second long. Once *tick()* has been called after the delay has elapsed, the event is sent with *sendNow()*.

Machine.java

aip.fsm.Machine The *Machine* class contains the states and transitions that define a FSM. It receives events sent from the *EventQueue*. It may have a *Context* for its actions or it may rely on the content of the events.

A summary of *Machine* is given in Table 4-4.

Machine(String name)
Machine(String name, Context context) Machines must have a name, since this is how they are identified and stored. The context, however, is optional. The context is used as a global data store for the FSM's actions.

Table 4-4

Machine summary

```
Construction
        Machine(String name)
        Machine(String name, Context context)

State Management
    int  countState()
  State  newState(String name)
  State  newState(String name, Action action)
  State  addState(State state)
  State  getState(String name)
  State  removeState(String name)
  State  setStartState(String name)
  State  getCurrentState()

Processing
   void  reset()
boolean  event(Event event)
Context  getContext()
   void  setContext(Context context)
 String  getName()
```

countState()
newState(String name)
newState(String name, Action action)
addState(State state)
getState(String name)
removeState(String name) A machine consists of one or more states. These methods provide the usual set of operations to create and access the states in a machine.

setStartState(String name) When the machine is created it has a *null* start state. Until the client specifies the name of an actual starting state, the machine cannot function. The start state may be the name of any of the states in the machine.

getCurrentState() As the machine processes events it changes states. This method returns the name of the currently active state. This is the only state that receives messages for the machine, at least until the machine transitions to a different current state.

reset() Calling *reset()* replaces the current state with the start state, effectively restarting the state machine. Note that it does not affect the context, which may contain data that also needs to be reset by the application.

event(Event event) The *EventQueue* calls the machine's *event()* method when it has an event for it. The client will rarely, if ever, send an event directly to a machine.

If the event is valid for this machine its *execute()* method is called and the event is passed to the machine's current state.

Typically, the result of processing an event is that the machine transitions to a new state. The code that does this is shown here:

```
String new_state = null;
if ((event.getState() == null)
   || (event.getState().compareToIgnoreCase(m_state.getName()) == 0))
{
   event.execute(m_context);
   new_state = m_state.event(m_context, event);
}
//
//
if (new_state != null)
{
   m_state = getState(new_state);
   m_state.enter(m_context);
   return true;
}
return false;
```

getContext( )
setContext(Context context) The *Context* is an application-defined container for any information used by the state machine's application-specific events, actions, and conditions.

getName( ) Retrieve the machine's name.

Event.java

aip.fsm.Event Events contain triggering information coming in to the machine from the outside world. The default *Event* class holds nothing more than targeting and scheduling information; it carries no semantic information.

You must create at least one application-specific event class for use by your state machine. This event sub-class will carry any information needed by the transition conditions.

A summary of *Event* is given in Table 4-5.

Event( )
Event(String machine) An event may be generic, to be received by the default *Machine* in the *EventQueue*, or it may be targeted at a specific

Table 4-5

Event summary

```
Construction
                    Event()
                    Event(String machine)

Processing
              void  execute(Context context)

Get/Set
final String  getMachine()
final String  getState()
  final long  getTime()
  final void  setMachine(String machine)
  final void  setState(String state)
  final void  setTime(long time)
```

machine. If a *Machine* name is specified the *EventQueue* searches for that machine and sends the event to it.

execute(Context context) The *execute()* method for events is rarely used. It is called just before the event is sent to the active state of the receiving machine.

get/set Methods An event may name a specific machine to receive it, or if the machine name is null it gets sent to the default machine. Likewise, it can name a specific *State* in that machine to receive it. If the destination state is not specified, the current state receives the event. Finally, an event may have a time associated with it so that the event can be processed with a delay.

The get/set methods provide access to these core *Event* attributes.

Context.java

aip.fsm.Context The *Context* interface is entirely empty. The FSM library makes no assumptions about your context.

There are three interfaces that must be filled in by the client application, since they interact together. These are *Context*, *Condition*, and *Action*.

State.java

aip.fsm.State States are what state machines are about. The rest of the classes are only here to support the states.

A *State*, at the minimum, consists of one or more *Transition* arcs that lead to different states. A *State* may also have an associated *Action* that is executed when the machine enters this state.

A summary of *State* is given in Table 4-6.

Table 4-6

State summary

Construction	
	`State(String name)`
	`State(String name, Action action)`
Transition Management	
`int`	`countTransition()`
`Transition`	`newTransition(Condition trigger, Action action, String state)`
`Transition`	`addTransition(Transition arc)`
`Transition`	`getTransition(int idx)`
`Transition`	`removeTransition(int idx)`
Processing	
`String`	`event(Context context, Event event)`
`void`	`enter(Context context)`
`String`	`getName()`

State(String name)
State(String name, Action action) All states need a name since this is how they are stored and accessed. A state may also have an optional *Action* that is executed when the state is entered.

newTransition(Condition trigger, Action action, String state) The *Transition* is the most prominent feature of a state. When a state receives an event it evaluates all of its transitions against the event and the first one that returns *true* is executed. Each *Transition* must have a *Condition* to trigger it and a destination *State* that the transition points to. After all, we are not transitioning if we do not go somewhere, even if it is back to the same state.

The optional *Action* is executed when the *Transition* is taken.

countTransition()
addTransition(Transition arc)
getTransition(int idx)
removeTransition(int idx) These methods provide the remaining support for manipulating the transition list. Of all the significant elements of the state machine the *Transition* is the only one that does not have a name. Since they are unnamed they are referenced by their index number in the list, starting with zero.

The *newTransition()* and *addTransition()* methods return the index number instead of the *Transition* object. Use *getTransition()* to get the actual object.

event(Context context, Event event) The *event()* method checks the *Event* against all of the transitions. If a transition returns true from *isValid()*, then that transition is followed.

```
int num = countTransition();
for (int idx=0; idx<num; idx++)
{
    Transition arc = getTransition(idx);
    if (arc.isValid(context, event))
    {
        return arc.execute(context, event);
    }
}
```

enter(Context context) When the state machine transitions to a state the *enter()* method of that state is executed. This runs the state's *Action*, if it has one.

Transition.java

aip.fsm.Transition Transitions make up the bulk of a state. They provide the pathways from one state to another and provide a place to attach actions.

A *Transition* has a *Condition* that determines if the transition is valid. When a *Transition* is executed it executes its *Action,* if it has one, and returns the name of its destination state.

A summary of *Transition* is given in Table 4-7.

Transition(Condition trigger, Action action, String state) A *Transition* consists of a *Condition* that determines when it is valid, a *State* that is the destination for the transition, and an optional *Action* to perform during the transition. These are all specified in the constructor.

Table 4-7

Transition summary

Construction	
	`Transition (Condition trigger, Action action, String state)`
Processing	
`boolean`	`isValid(Context context, Event event)`
`String`	`execute(Context context, Event event)`
Get/Set	
`String`	`getDestination()`
`Condition`	`getCondition()`
`Action`	`getAction()`
`void`	`setDestination(String state)`
`void`	`setCondition(Condition trigger)`
`void`	`setAction(Action action)`

isValid(Context context, Event event) This method tests the validity of the transition's *Condition*. It passes the *Context* and *Event* to the condition's *evaluate()* method and returns the result. Note that the condition is known as *m_trigger* inside the *Transition*.

```
if (m_trigger == null)
{ return false; }

return m_trigger.evaluate(context, event);
```

execute(Context context, Event event) Execute any action associated with the transition and return the destination state.

```
if (m_action != null)
{ m_action.execute(context, event); }

return m_dest;
```

get/set Methods A transition consists of a condition, destination, and action, so these attributes are accessible through the usual get and set methods.

Action.java

aip.fsm.Action The behavior of the state machine lies in the *Action* objects attached to it. Since the actions are application specific, the *Action* interface is minimal. When implementing an *Action* you only have to specify an *execute()* method.

```
public void execute(
    Context context,///< Context we execute against
    Event event );/// < Event we execute against
```

Note that the *execute()* has access to both the machine's *Context* and the current *Event*. Actions attached to a state have a *null event* parameter.

Condition.java

aip.fsm.Condition Conditions are used to guard the gateways to the transitions. When a *Condition* evaluates to *true*, for whatever reason, its *Transition* is allowed to execute.

Like *Action*, *Conditions* are application specific. The subclasses that implement *Condition* only have to provide an evaluator.

```
public boolean evaluate(
    Context context,/// < Context we evaluate against
    Event event );/// < Event we evaluate against
```

A condition has access to the context and event, so it can make an informed decision. There will always be an event to test, unlike the *Action* where it may be null.

TrueCondition.java

aip.fsm.TrueCondition A *TrueCondition* is an implementation of *Condition* that returns *true* as its evaluation answer. This condition is useful for making a default transition for a state.

ActionList.java

aip.fsm.ActionList An *ActionList* is a list of one or more *Action* objects. When the *ActionList* is executed it walks through its list of actions and executes them.

```
int num = m_actlist.size();
for (int idx=0; idx<num; idx++)
{
   Action act = (Action)m_actlist.get(idx);
   act.execute(context, event);
}
```

The *ActionList* is used when you want to do something complicated during a transition. Instead of making one complicated *Action* you can tie a bunch of simpler actions together in a list.

A summary of *ActionList* is given in Table 4-8.

EventAction.java

aip.fsm.EventAction The *EventAction* is an interesting implementation of the *Action* interface. It provides a way for an *Action* to send an *Event* to a *Machine*. It also demonstrates how you can access the *EventQueue* in an action.

A summary of *EventAction* is given in Table 4-9.

EventAction(Event event)
EventAction(Event event, long delay) The two constructors for *EventAction* reflect the two ways events can be sent. The one form sends events immediately upon execution and the other form sends the events with a time delay.

Table 4-8 ActionList summary

Construction	
	`ActionList(Action act)`
`Void`	`add(Action act)`
Processing	
`void`	`execute(Context context, Event event)`

Table 4-9

EventAction summary

Construction

```
    EventAction (Event event)
    EventAction (Event event, long delay)
```

Processing

```
void execute(Context context, Event event)
```

execute(Context context, Event event) The execution of this action gets the *EventQueue* instance and sends it an event.

```
EventQueue queue = EventQueue.getInstance();

if (m_delay < 1)
{ queue.sendNow(m_event); }
else
{ queue.sendDelay(m_event, m_delay); }
```

Note that similar actions can be created to set the default machine for the queue or perform other actions outside of the context.

Code: Coin Box Revisited

We have added more capabilities to our FSM since the last coin box implementation. One of the more intriguing features is the delayed message. This, combined with the ability for a message to target a particular state, can provide feedback loops.

One use for timed messages that I have seen is that of the rocket from the game Quake (ai-depot.com/FiniteStateMachines/FSM.html). When the rocket is fired it flies through the air until it either hits something or times out and goes away. This might look like Figure 4-7.

Most of the rocket's behavior is managed outside of the state machine, based on the state the machine is in. During *Spawn* the rocket launching animation occurs. When that is complete the *done spawning* event moves the machine into the *Move* state. During the transition to *Move* a time-delay event is sent to *Move*. During the *Move* phase the physics and animation systems guide the rocket. Eventually the rocket will receive the *timeout* event or it will hit something and *Explode*. Either way it goes to *Die* and is removed from the game.

We can apply a similar timeout event for the coin box machine used earlier. A simplified diagram for the new and expanded coin box is given in Figure 4-8.

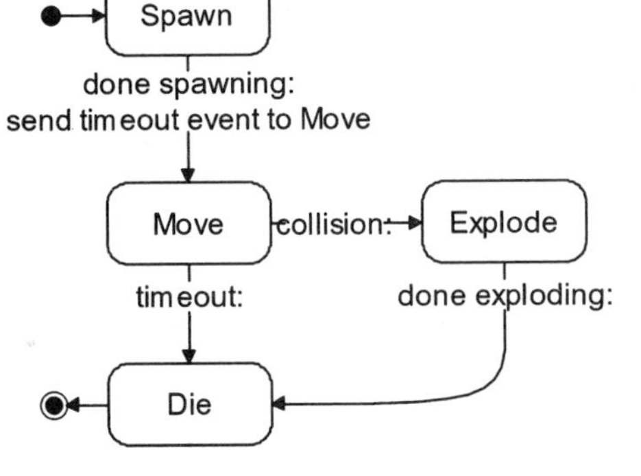

Figure 4-7 Possible rocket FSM

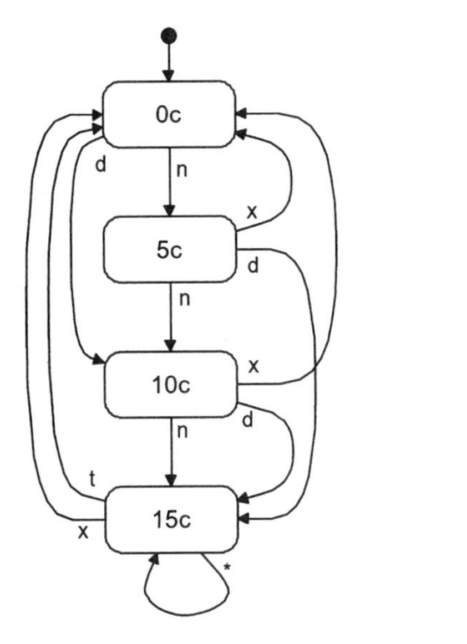

Figure 4-8 Expanded coin box FSM

The diagram is getting cluttered! To alleviate the pain it only shows the events and not the actions associated with them.

The meaning of the event tags are:

- n Nickel was inserted
- d Dime was inserted
- x Timeout
- t Take product from the machine
- * Default (matches all events)

The details of the machine are spelled out in detail when we discuss the *FSMTest* code.

In addition to the test application *FSMTest*, there are four application-specific classes that extend the basic FSM library: *CharAction*, *CharCondition, CharCondition*, and *CharEvent.*

The expanded class diagram is shown in Figure 4-9.

CharEvent.java

aip.app.fsmtest.CharEvent The coin box state machine operates on characters. The input to a machine could be anything, however, in this case characters are the easiest. A *CharEvent* is simply an implementation of *Event* that holds a single *char*.

A summary of *CharEvent* is given in Table 4-10.

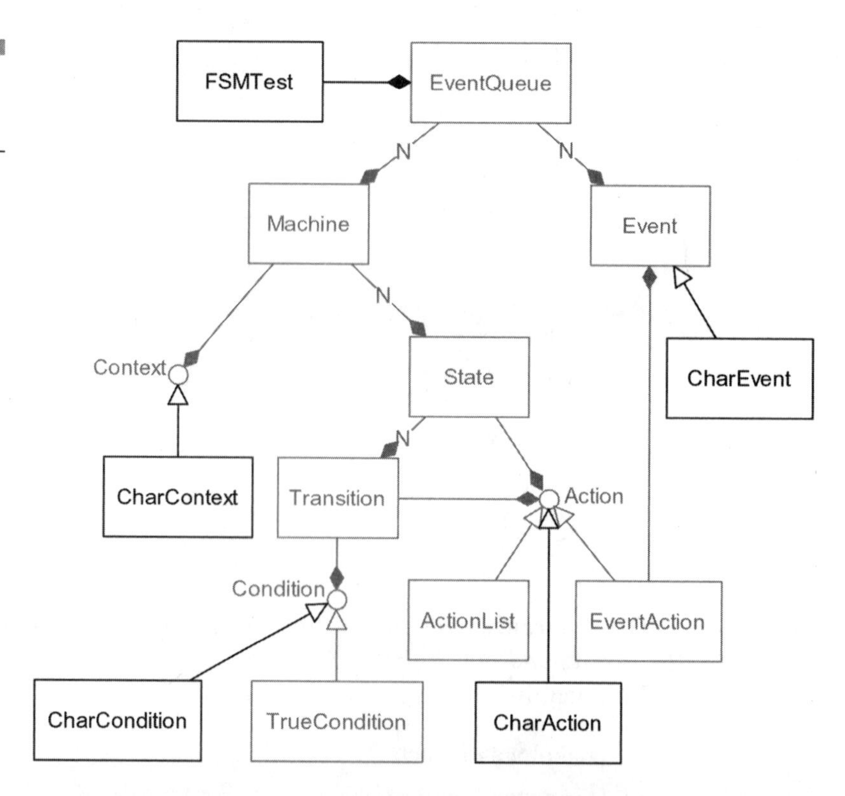

Figure 4-9 Expanded coin box class diagram

Table 4-10 CharEvent summary

Construction

```
CharEvent(char token)
CharEvent(char token, String machine)
```

Get/Set

```
char getChar()
void setChar(char token)
```

Table 4-11

CharContext summary

Public Methods	
	`CharContext()`
`void`	`set(char in)`
`char`	`get()`

CharContext.java

aip.app.fsmtest.CharContext The *CharContext* class is not actually used by *FSMTest*, but it provides an example of creating a context that is consistent with the needs of the application.

A summary of *CharContext* is given in Table 4-11.

CharCondition.java

aip.app.fsmtest.CharCondition *CharCondition* performs tests on the character held by a *CharEvent*. Note that the *CharContext* is not used. *CharCondition* may evaluate the character with tests like *IS_ALPHA* and *IS_NUM*, or it can compare it to another character with tests like *EQUAL* and *GREATER*.

Tests are defined as *CharCondition* objects and the testing is done with *evaluate()*.

A summary of *CharCondition* is given in Table 4-12.

CharAction.java

aip.app.fsmtest.CharAction The *CharAction* class has two forms. In one form the action prints (emits) a fixed string to the console. In the other form it echoes or manipulates the character in the *CharEvent*. In both forms the *CharAction* constructor is passed a *JTextArea,* where the results of the actions are displayed.

In hindsight, the *JTextArea* could be a part of the *CharContext* and then *CharAction* could print to that context, which would then echo to the text area.

A summary of *CharAction* is given in Table 4-13.

FSMTest.java

aip.app.fsmtest.FSMTest *FSMTest* is, as its name suggests, a FSM test application. It creates the state machine outlined in Figure 4-9 and then feeds it events in the form of characters entered in its UI.

Part of *FSMTest* is devoted to setting up the user interface and then processing characters. However, the significant method is *build_coinbox(),* which constructs the state machine.

A summary of *FSMTest* is given in Table 4-14.

Table 4-12

CharCondition summary

```
Construction
            CharCondition(int test)
            CharCondition(char token, int test)

Processing
  boolean   evaluate(Context context, Event event)

Static Public Attributes
final int   EQUAL = 0
final int   NOT_EQUAL = 1
final int   GREATER = 2
final int   GREATER_EQUAL = 3
final int   LESS = 4
final int   LESS_EQUAL = 5
final int   EQ = EQUAL
final int   NE = NOT_EQUAL
final int   GT = GREATER
final int   GE = GREATER_EQUAL
final int   LT = LESS
final int   LE = LESS_EQUAL
final int   IS_ALPHA = 6
final int   IS_NUM = 7
final int   IS_ALNUM = 8
final int   IS_SPACE = 9
final int   NOT_ALPHA = 10
final int   NOT_NUM = 11
final int   NOT_ALNUM = 12
final int   NOT_SPACE = 13
```

Table 4-13

CharAction summary

```
Construction
            CharAction (JTextArea display, int act)
            CharAction (JTextArea display, String out)

Processing
     void   execute (Context context, Event event)

Static Public Attributes
final int   SKIP = 0
final int   ECHO = 1
final int   EMIT = 2
final int   UPPER = 4
final int   LOWER = 5
```

build_coinbox() First it creates a machine in the *EventQueue* and sets it as the default.

```
Machine machine = m_fsm.newMachine("CoinBox");
m_fsm.setDefault("CoinBox");
```

Table 4-14

FSMTest summary

Construction	
	`FSMTest()`
`void`	`assemble()`
State Machine	
`void`	`build_coinbox()`
`void`	`run()`
UI Interaction	
`void`	`actionPerformed(ActionEvent e)`
`void`	`keyTyped(KeyEvent e)`

With that done, it creates the various conditions that guard the arcs between states.

```
CharCondition dime = new CharCondition('d', CharCondition.EQUAL);
CharCondition nickel = new CharCondition('n', CharCondition.EQUAL);
CharCondition take = new CharCondition('t', CharCondition.EQUAL);
CharCondition timeout = new CharCondition('x', CharCondition.EQUAL);
```

There are several actions to define. In this character-based example each action is a string for the text area indicating that someone inserted a nickel or dime, that a timeout occurred, that the machine is giving change, and so on.

```
CharAction dime_action = new CharAction(m_output, "(dime)");
CharAction nickel_action = new CharAction(m_output, "(nickel)");
CharAction change_action = new CharAction(m_output, "Give 5c
Change");
CharAction take_action = new CharAction(m_output, "(take)");
CharAction reject_action = new CharAction(m_output, "(reject
coin(s))");
CharAction open_action = new CharAction(m_output, "Door Open");
CharAction close_action = new CharAction(m_output, "Door Close");
```

A special delayed *EventAction* is created, so the machine can timeout. In this case, it times out when 5 seconds have elapsed since the first coin insertion.

```
CharEvent timeout_event = new CharEvent('x', "CoinBox");
EventAction start_timeout = new EventAction(timeout_event, 5000);
```

When the first coin is inserted the transition executes both the coin's generic event and the delayed timeout event. This is described using an *ActionList*.

```
ActionList first_dime = new ActionList(dime_action);
first_dime.add(start_timeout);
```

```
ActionList first_nickel = new ActionList(nickel_action);
first_nickel.add(start_timeout);
```

Other action lists include opening the box while giving change and closing the box while giving change.

```
ActionList open_change = new ActionList(open_action);
open_change.add(change_action);

ActionList timeout_action = new ActionList(reject_action);
timeout_action.add(close_action);
```

Finally, the states are created and the transitions are attached to them. The transitions are defined using the conditions and actions specified earlier.

```
State state = machine.newState("0c");
state.newTransition(dime, first_dime, "10c");
state.newTransition(nickel, first_nickel, "5c");

state = machine.newState("5c");
state.newTransition(dime, dime_action, "15c");
state.newTransition(nickel, nickel_action, "10c");
state.newTransition(timeout, timeout_action, "0c");

state = machine.newState("10c");
state.newTransition(dime, open_change, "15c");
state.newTransition(nickel, nickel_action, "15c");
state.newTransition(timeout, timeout_action, "0c");

state = machine.newState("15c", open_action);
state.newTransition(take, take_action, "0c");
state.newTransition(timeout, timeout_action, "0c");
state.newTransition(new TrueCondition(), reject_action, "15c");
```

Now that the machine is built we can indicate its start state and begin passing events to it.

```
machine.setStartState("0c");
```

keyTyped(KeyEvent e) This takes the character from Java's *KeyEvent* and passes it to the FSM *EventQueue* as a *CharEvent*.

```
public void keyTyped(KeyEvent e)
{
m_fsm.sendNow(new CharEvent(e.getKeyChar(), null));
}
```

A sample run of the machine is shown in Figure 4-10. Note that the last cycle of the machine accepted two nickels and then timed out.

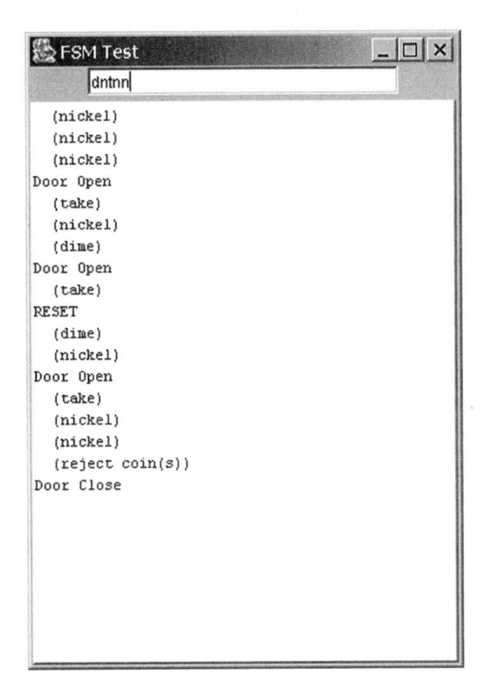

Figure 4-10
Coin box

Markov Models

The Markov model is used to describe the probability of one state following another. The Markov property says that the probability of a state occurring depends only on the current state and not any additional history of the transaction.

In effect, a Markov model is a probabilistic state machine.

The probabilities are typically found by analyzing sample input data. They could also be set by hand using your assumptions about the data.

For example, let us look at a model built from sample data from Allen (1995). This sample corpus consisted of 1,998 words identified to be in four categories, N, V, Art, and P. This is an unnaturally small corpus and set of categories, but it provides a tractable data set.

The words relevant to this discussion and their use frequencies are given in Table 4-15.

The probabilities of the transitions between categories are listed in Table 4-16 and the resulting state machine is shown in Figure 4-11.

Once you have such a model there are several things you can do with it. You can calculate the probability of a given sequence, you can

Table 4-15

Sample words and categories

	N	V	Art	P	Total
flies	21	23	0	0	**44**
like	10	30	0	21	**61**
a	1	0	201	0	**202**
flower	53	15	0	0	**68**
bird	64	1	0	0	**65**
other	684	231	357	286	**1,558**
Total	**833**	**300**	**558**	**307**	**1,998**

Table 4-16

Transition probabilities

Current state	Next state	Probability
null	N	0.29
	V	0.71
N	N	0.13
	V	0.43
	P	0.44
V	N	0.35
	Art	0.65
Art	N	1.00
P	N	0.26
	Art	0.74

determine the probable states in a hidden Markov model, or you can generate sequences in the style of the sample corpus.

More information on Markov models is easily found in books like Allen (1995) and Ballard (1997).

Probability of a Sequence

Given a sequence of states (inputs), you can determine the probability of that sequence having occurred. This is done by accumulating the

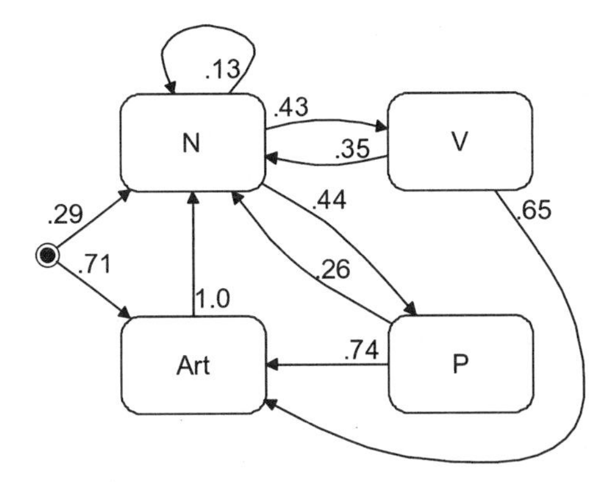

Figure 4-11 Markov state machine

probabilities of the each transition through the sequence. For example, given the sequence:

N V Art N

The probability of this sequence is given as:

$$p(\phi,N) * p(N,V) * p(V,Art) * p(Art,N)$$
$$0.29 * 0.43 * 0.65 * 1.0$$
$$= 0.081 \quad \textbf{4-1}$$

where *p(A,B)* is the probability of making the transition from state *A* to state *B*.

Probable States in a Hidden Markov Model We do not always know the actual state a modeled system is in but must try to deduce it from its output. Using this output we can determine the *likely* states that were followed by the process we are modeling, but we will not know for sure. When the states of the model are hidden, the model that tries to predict the states is a Hidden Markov Model, or HMM.

For example, in the sentence "Flies like a flower" each word is in one of several possible parts of speech. Since we do not know the parts of speech intended (the state is hidden) we can only attempt to discover the probable part of speech for each word.

With four words and four categories, there are 4^4 or 256 possible sentences to process and check the probabilities. Of course, for longer sentences and more categories, a comprehensive search quickly becomes impractical.

While there are different ways to process a Markov model, one technique to simplify this problem is the Viterbi algorithm.

In the Viterbi algorithm, you step through the input sentence and accumulate the transition probabilities for the word in each category. After each step, use the highest scoring version from each category as the basis for the next step.

This is much easier to illustrate than explain, so let us walk through our sample sentence and see how it works. Using the sentence:

Flies like a flower

The first word is "Flies" and we do not have a state, so we begin with the probability of "flies" being used as the first word.

There are different ways to calculate this probability. One way is to use a combination of a word in the probability of the category *N* being the specific word *flies* and the probability of category *N* beginning a sentence:

$$\begin{aligned} &p(\mathrm{N} \mid \mathrm{flies}) * p(\phi, \mathrm{N}) \\ &0.025 * 0.290 \\ &= 0.00725 \end{aligned} \quad \textbf{4-2}$$

where:

- $p(\mathrm{B} \mid \mathrm{A})$ is the probability of B given A
- $p(\mathrm{A}, \mathrm{B})$ is the probability of the transition from A to B

or you could use the transition probabilities from the model directly. One problem with this is that a state with only one transition, such as Art to N at 1.0, can force that line of inquiry to win because the total probability never goes down.

The probability we use is similar to Equation 4-2:

$$\begin{aligned} &p(\mathrm{flies} \mid \mathrm{N}) * p(\phi, \mathrm{N}) \\ &0.477 * 0.290 \\ &= 0.1383 \end{aligned} \quad \textbf{4-3}$$

which is the probability that "flies" is in category N and the probability that category N starts the sentence.

The probabilities for "flies" as the first word in its two legal categories are shown in Table 4-17. The other categories, Art and P, are not legal at the start of the sentence or for the word "flies".

The only states with valid probabilities are flies/N and flies/V, so these are used as the root of the next step, shown in Table 4-18.

Table 04-17

Start to flies

State probability	State	Arc probability	Destination state	Total probability
	ϕ	0.1383	flies/N	0.1383
		0.3713	flies/V	0.3713

Table 04-18

flies to like

State probability	State	Arc probability	Destination state	Total probability
0.1383	flies/N	0.0213	like/N	0.0029
		0.2116	***like/V***	***0.0293***
		0.0	like/Art	0.0
		0.1514	like/P	0.0209
0.71	flies/N	0.0574	***like/N***	***0.0408***
		0.0	like/V	0.0
		0.0	like/Art	0.0
		0.0	like/P	0.0

Table 04-19

like to a

State probability	State	Arc probability	Destination state	Total probability
0.0408	like/N	$6.50*10^{-4}$	***a/N***	***$2.652*10^{-5}$***
0.0293	like/V	$1.75*10^{-4}$	a/N	$5.127*10^{-6}$
		0.6468	**a/ART**	**0.0190**

The highest scoring categories, one each for N, V, etc, are used as the roots of the next step. Ignoring states and transitions with zero probability, the next step is shown in Table 4-19.

The last step is shown in Table 4-20.

The final winning state is flower/N with a probability of 0.0735. Tracing backwards from this we can see that the most probable states for the sentence are :

Flies/N like/V a/Art flower/N

which is quite reasonable.

An interesting exercise would be to perform this analysis on the sentence “flies like a bird”.

Table 04-20
a to flower

State probability	State	Arc probability	Destination state	Total probability
$2.653*10^{-5}$	a/N	0.6202	flower/N	$1.664*10^{-6}$
		0.2249	flower/V	$5.966*10^{-6}$
0.0944	a/Art	0.779	***flower/N***	***0.0735***

Code: Markov Sentence Generation

Using a Markov model in the form of a state machine we can generate sentences that follow the probabilities of the model. Of course, a "sentence" can be any series of outputs such as game or robot behaviors and not necessarily words.

Beginning at some start state the machine takes a random arc to the next state. Each arc is followed with a probability determined by the Markov model.

There are several processes in this example. The first step reads in one or more text files and parses them into individual words. Note that punctuation marks are considered "words".

The parsed words are then sent to a Markov context that keeps track of the transitions from one state to the next.

This context is then converted into a state machine and that state machine is processed to generate output with similar statistical properties as the input.

There are many different orders of statistical generation.

A zero-order process generates states completely randomly, without reference to the model's word or transition distribution.

A first-order process generates states based on the word use probabilities in the sample corpus.

A second-order process is the first process that can be considered a Markov model. At this level there is a reference state providing a context for the destination state. Transitions from this reference state to the destination state are determined by the transition probabilities established by the corpus.

A third-order process uses a two-state context from the corpus to find the probable next state, and so on.

Because it is both short and distinctive we use the poem *Jabberwocky* by Lewis Carrol (from the book *Through the Looking Glass and What Alice Found There*, 1872) as our sample text. This poem is reproduced in full in Table 4-21.

Table 4-21
Jabberwocky

Jabberwocky
by Lewis Carroll

`Twas brillig, and the slithy toves
Did gyre and gimble in the wabe:
All mimsy were the borogoves,
And the mome raths outgrabe.

"Beware the Jabberwock, my son!
The jaws that bite, the claws that catch!
Beware the Jubjub bird, and shun
The frumious Bandersnatch!"

He took his vorpal sword in hand:
Long time the manxome foe he sought —
So rested he by the Tumtum tree,
And stood awhile in thought.

And, as in uffish thought he stood,
The Jabberwock, with eyes of flame,
Came whiffling through the tulgey wood,
And burbled as it came!

One, two! One, two! And through and through
The vorpal blade went snicker-snack!
He left it dead, and with its head
He went galumphing back.

"And, has thou slain the Jabberwock?
Come to my arms, my beamish boy!
O frabjous day! Callooh! Callay!'
He chortled in his joy.

`Twas brillig, and the slithy toves
Did gyre and gimble in the wabe;
All mimsy were the borogoves,
And the mome raths outgrabe.

The project is called *Markov*. There are four new classes extending the state machine: *MarkovContext* with its the sub-class *MarkovLink*, *ProbCondition*, and *ProbEvent*.

MarkovContext.java

aip.app.markov.MarkovContext The *MarkovContext* provides a specialized state machine context that records the words and transitions between words from the source corpus. It counts each use of a word and transitions from that word, so the transition probabilities can be generated.

aip.app.markov.MarkovLink *MarkovContext* contains the local sub-class *MarkovLink*. This class is used to record a transition from one word to any following words.

ProbEvent.java

aip.app.markov.ProbEvent The probability event generates a new random number between 0 and 1 with each use. It is one of the few events that make use of the event *execute()* method.

ProbCondition.java

aip.app.markov.ProbCondition The probability condition is a condition that evaluates to true only when the value in the probability event falls between a specific range of values. The size of this range determines the probability that this condition will be true.

```
double rnd = ((ProbEvent)event).getProb();
if ( (rnd >= m_low)
   && (rnd < m_high) )
{
   return true;
}
return false;
```

Markov.java

aip.app.markov.Markov The *Markov* application creates a state machine to parse the input files. This machine then feeds the *MarkovContext* that collects statistics about the input files.

Using the data in the context, *Markov* creates a second state machine that models the collected statistics.

Finally, *Markov* executes the Markov machine to generate output in the style of the input corpus.

A summary of *Markov* is given in Table 4-22.

Table 4-22 Markov summary

Public Methods	
	Markov (int order)
void	buildParser ()
void	parse (String filename)
void	buildGenerator ()
void	generate (int num)
void	statistics ()
Static Public Methods	
void	main (String args[])

Results This is a command-line application, so you execute it with a command like this:

```
java aip.app.markov.Markov 2 50 jabberwocky.txt
```

The first parameter is the order of the machine to run, the second number is the size of output to generate, and all remaining parameters list files to analyze.

The results shown below are from runs of the *Markov* program. Since the output system of that program is naïve, capitalization and spaces have been adjusted for easy readability.

A second-order machine uses one word of context, such as "twas" leading to "brillig". The second-order machine generates results like that shown in Table 4-23.

A third-order machine uses two words of context, giving more legible results as shown in Table 4-24. This builds links like "twas brillig" leading to *comma*.

A fourth-order machine really limits the choices available for the transitions, especially with a sample as small as *Jabberwocky*. Fourth-order links look like "did gyre and" leading to "gimble". Fourth-order results are shown in Table 4-25.

Even though each order of Markov machine produces better output, it is still definitely babbling. There is no underlying meaning or intent in the output so it is nothing more than an interesting curiosity.

In domains other than human language Markov-generated outputs may be perfectly acceptable.

Table 4-23

Second-order output

All mimsy were the tulgey wood, two! And gimble in his vorpal sword in his vorpal sword in hand: long time the wabe; all mimsy were the wabe: all mimsy were the slithy toves did gyre and through and burbled as in his vorpal blade

Table 4-24

Third-order output

Jaws that bite, The jabberwock? Come to my arms, my beamish boy! O frabjous day! Callooh! Callay! He left it dead, and with its head he went galumphing back. And, has thou slain the jabberwock, with eyes of flame,

Table 4-25

Fourth-order output

And burbled as it came! One , two! And through and through the vorpal blade went snicker - snack! He left it dead, and with its head he went galumphing back. And, as in uffish thought he stood, the jabberwock, my son! The

One way a Markov system could be used is to try and predict a user's actions in some environment. For example, games tend to have simplified worlds with a restricted range of actions. The transitions in an N-order Markov machine could be calculated based on how a player interacts with the world. Once a reasonable amount of history has been accumulated, the computer could get a sense of what the player might do next in a situation.

Another, similar, use is to take these same user statistics and the computer can play in the style of the trainer.

The system developed in this chapter does not lend itself to online, interactive learning. It reads all of the input in a batch and then creates and runs the machine as a separate step. A learning system would need to blend the input and the output into one continuous process where the Markov system adapts its links over time.

Frame-Based Intelligence and Chatbots

Time to swim up from the depths of Java code and catch some air. This section explores some other ways that you can use data to drive an artificial intelligence.

The oldest computer program designed to hold a conversation with a human is Eliza, written at MIT by Joseph Weizenbaum. It was first described in the journal *Communications of the ACM* in January 1966. The type of conversation it simulated was very specific, that of a Rogerian analyst working with a patient.

This program worked well, sometimes convincing people that they were conversing with another person. And what technology, what great AI system did Eliza use? None. It cheated.

Cheating, of course, has a long and illustrious history when it comes to computer AI. Probably because the "real thing" is so amazingly hard. But how do you make the distinction between a clever "cheat" and a truly intelligent machine?

One way of looking at computer AI is through the Chinese Room analogy, defined by John Searle in 1990 as part of his argument of why AI will never be "really" intelligent.

Take someone who speaks fluent Chinese, talking to another person through a terminal. Now remove the person on the other side of the conversation. The terminal now spits out cards with symbols on them. A team of people in a big room take these cards, cross-reference them against rulebooks, compile, organize, sift, and process them. In the end, they have another stack of cards that they feed into a machine that transcribes them back to the human side of the conversation.

Maybe the team in the room is blindingly fast. Or maybe this is a conversation through the mail.

The people in the room at no time have any understanding of Chinese and the human talking to them cannot tell they are not talking to another human. Is the process in the room intelligent? Why? Why not?

Of course the room full of cards and people is simply an analogy for what happens inside a computer.

Alan Turing understood the difficulty of knowing whether something was intelligent. He came up with a definition of intelligence that is much like the definition of obscenity; people recognize it when they see it.

In 1950 Turing turned the question "can a computer think?" around and asked instead, "if a computer could think, how could we tell?" Turing introduced his now-famous test for intelligence. If the computer can fool a human into believing that it, too, is human, it should be considered intelligent. Needless to say, this proposal has generated 50 years of controversy.

Forty years later Hugh Loebner formalized the Turing Test into a challenge for programmers. He offered a grand prize of $100,000 to the first computer to successfully imitate a human. Until that prize is claimed there is a lesser prize of $2,000 for the best attempt during the annual competition.

The programs entered into the competition so far have not been serious attempts at deep intelligence, but instead aim to win by cheating.

Which leads us back to Eliza.

Pattern Matching

Eliza operates on the simple principle of keyword (or key phrase) matching. The user types in some sentence and, after appropriate parsing and cleaning up, the pattern matcher takes over. Perhaps it sees the words "I" and "remember" and then more words, perhaps "the day I broke my leg". Aha! That matches this rule, taken from Charles Hayden's Eliza Test script:

```
key: remember 5
 decomp: * i remember *
 reasmb: Do you often think of (2) ?
 reasmb: Does thinking of (2) bring anything else to mind ?
 reasmb: What else do you recollect ?
 reasmb: Why do you remember (2) just now ?
 reasmb: What in the present situation reminds you of (2) ?
 reasmb: What is the connection between me and (2) ?
  reasmb: What else does (2) remind you of ?
```

In the response templates any words before the “I remember” key words are represented by (1) and any words after them are (2). This provides the system with several response choices. Perhaps this time it chooses the second and types back:

```
Does thinking of the day you broke your leg bring anything else to
mind?
```

Note that it conjugated the relevant words. It helps to cheat if you have a basic grasp of grammar.

Even with the clever word transformations, this is just an exercise in template matching. The next time “I remember” comes up a clever program would pick a different response, until it is forced to repeat itself.

This type of pattern matching conversation program is commonly known as a “chatbot”, and Simon Laven provides handy links to many types at www.simonlaven.com. Another place where you can learn about bots, chat and otherwise, is at www.botknowledge.com. Bots are a popular topic and are being applied to customer support, search engines, and other places where the computer needs to interact with people.

One chatbot is A.L.I.C.E., from the ALICE AI Foundation. This bot is notable for two reasons. One is that it won the Loebner prize in 2000 and 2001. The other thing about ALICE is that its technology is open-source under the GNU General Public License, with a number of implementations available. Find out more at www.alicebot.com.

In addition to matching rules, chatbots can use induction rules that resolve against the user’s input, dynamic Markov models that learn as they are used, and so on. Many techniques have been applied to chatbots, but most of them follow the basic format of finding a key word or phrase and then responding with a pre-defined or statistically generated sentence.

Regardless of their technology, chatbots create the illusion of intelligence through the clever design of their internal scripts. The intelligence is actually a reflection of their designer and not based on deeper understanding.

Scripts

The pattern matching and responses in the world of the chatbot are limited to the statements and responses of language. That system could be extended with additional capabilities.

In a more diverse system, picking on games again, the script could provide a multimedia response. Given text or behavior as input from the user the response generator could type or speak a reply, trigger sound effects, start an animation, or perform any other action available in the environment.

These scripts can be specialized programming languages dedicated to their application. Little languages and scripting systems are a large topic by themselves. The interested reader should have no problem finding references on the subject.

Frames and Agendas

The chatbot matching rules provide a framework where the program can hang the user's input. They have three slots, one of which is the key being matched:

<pre> key <post>

These slots consist of everything typed before the key, the key phrase, and then everything after the key. With appropriate grammatical wizardry, this information can be transformed into a credible response.

However, much more is possible. With a little extra effort a partial parser could be written that extracts verb and noun phrases from the sentence. These can then be used to fill the slots of a more intricate frame (Allen, 1995).

In the boring yet practical domain of airline reservations the slots could involve the information needed to book a flight:

- Flight Number
- Origin City
- Destination City
- Departure Date
- Departure Time
- Return Date
- Return Time

- Layover City
- Layover Delay

The user would initiate the conversation with an input like:

I want to book a flight to Boston.

This sets the *Destination City* field to *Boston,* and it might also set the *Origin City* field if the system could safely make that assumption. If it is embedded in a kiosk in the Austin airport it could assume the *Origin City* is *Austin*.

If this were a conversational chatbot it would look at the frame and treat it as a template to match against its internal store of clever responses.

For a goal-driven application, such as this kiosk that is trying to sell a ticket, the empty slots in the frame provide an agenda that drives the dialog. Perhaps the next thing it wants to learn is the date of departure, so it will generate the reply:

What day do you want to fly to Boston?

And so on, filling and refilling the slots based on user input until the frame is full. It can then finish the dialog with the final question:

What is your credit card number?

Some of the slots in the frame are not filled by the user but by the database. *Flight number* and the *Layover* information, for example. If the user does specify that they want to travel through London, the system should add that constraint to the frame.

This is still cheating. The computer has no concept of what an airplane is or that Austin is a great place to listen to music. It is effective cheating since it provides a smooth, intelligent-seeming interface to achieve its goal.

And that is what really matters.

CHAPTER 5

Discrete Searching

This chapter introduces search trees and graphs and looks at techniques for finding things within them.

Since many problems are too big for exhaustive searches, we use the A* search to find a path through the problem without examining every node of the graph. The versatile A* is applied to a physical path-finding problem and an abstract material utilization task.

Some problems consist of turn-taking, where each level in a search tree has a conflicting goal. The min–max search addresses these two-player "games" and alpha–beta pruning makes them efficient. The alpha–beta search is then applied to a simple game.

Additional search details and algorithms can be found in Bolc (1992).

Introduction

A discrete search looks for a match among a countable number of things. These "things" may not just be items in a list but could represent choices or states in a system of behavior. Discrete systems are also known as "table based", since the items can be listed in a table.

The simplest representation of these choices is as a list of options. If the list is sorted, it can be searched using a binary search. For more efficient insertion of items into a dynamic list it can be modeled as a binary tree (Figure 5-1). This tree gives us choices on how to wander through its nodes, some of which are explored in the section on brute-force searching.

The tree in Figure 5-1 consists of nodes (*A*, *B*, *C*) and branches (*1*, *2*). The terminal nodes (*D*, *E*, *F*, *G*) are also known as the leaves of the tree. The very top node (*A*) is the root.

The nodes and/or leaves can be objects in the tree or they can be states. States are placeholders that represent the choices taken up to that point. The branches may be meaningless links that hold the nodes together or they can be decisions in a process.

Figure 5-1
Binary tree

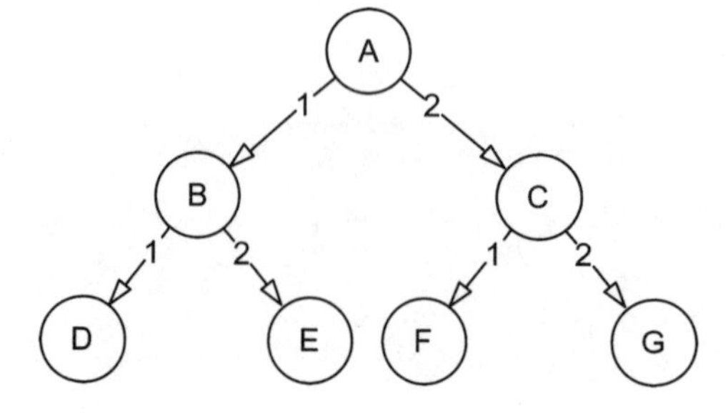

For example, node E may be an object with content E'. A search will be looking for that content and, when it finds it, the search can be done.

E may also represent the choice sequence [*1*, *2*] that leads to this state. A search will be looking for the best possible series of choices. All of the states are evaluated and the best path wins. The application can then make the choices that lead to the winning state.

Trees are not limited to the binary variety. The nodes can have any number of branches. In chess, for example, the root node could represent the current state of the pieces on the board. Each branch is a move that is one move away from the previous state. With all of the pieces and their various moves the tree quickly explodes into a staggering number of possibilities. For these types of trees the brute-force solutions do not work anymore and we drift in to the realm of AI.

Trees are a specific form of acyclic graph, such as that shown in Figure 5-2. In this graph both choices out of C lead to F. This situation is common when traversing a map or two-dimensional grid of choices. Technically a graph can be converted to a tree by duplicating the overlapping node.

Cyclic graphs take this branching and merging one step further. They allow branches back up the tree creating loops.

Many search methods do not work with cyclic graphs or acyclic graphs but require a tree. For searches on graphs it can be important to keep a list of visited, or closed, nodes. This way each node is only processed once and loops are avoided.

In game-like environments each layer of the tree or graph represents a different phase of turn-taking. Player 1 makes a move, then player 2, then back to 1, and so on. A specialized set of search methods exists for these turn-taking graphs.

In the abstract, a search is exploring the space of all possible states in order to find a state that best matches its goal. The match can be a data match or be based on a complex scoring scheme. The set of all possible configurations, or states, is called the state space.

Figure 5-2
Acyclic graph

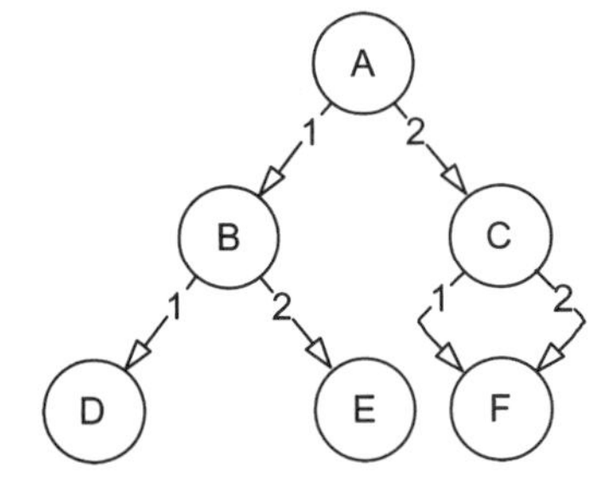

It can be difficult for the computer to know if one state arrangement, such as a particular arrangement of pieces in Go, is better than another. This is a different problem from the search itself and its solution is specific to the application at hand. Fortunately, the search algorithms do not care about how the decision is made, only that it returns a useable score.

Brute-Force Searching

Brute-force search techniques are simple and do not warrant a deep study here. We briefly look at these techniques for trees, avoiding the complexities of graphs.

These searches traverse a tree, trying to find a particular entry in that tree. If the search is looking for an exact match, it terminates when it finds it. If it is looking for the lowest or highest scoring leaf on the tree, based on some arbitrary criteria, it may have to look at each and every node in the tree.

Depth-First

Depth-first searching is perhaps the easiest form of search. It dives down to the bottom of the tree first, searching an entire branch before moving on to the next one. Depth-first search order is shown in Figure 5-3.

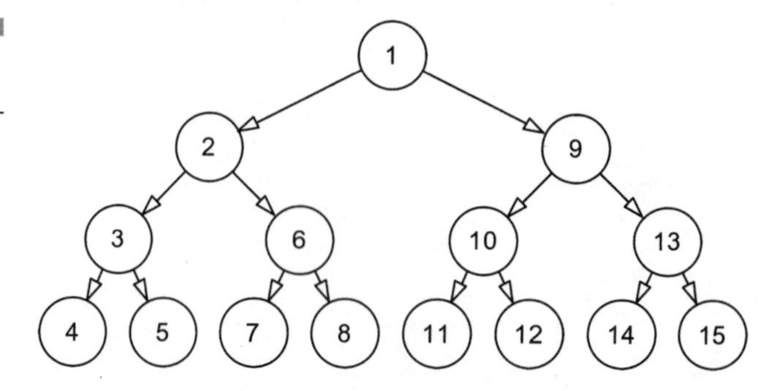

Figure 5-3
Depth-first search

This strategy is easy to implement recursively:

```
depthFirst(Node node)
{
    Iterator iter = node.iterator();
    while (iter.hasNext())
    {
        Node child = (Node)iter.next();
        //
        // If the child is the goal, quit; or otherwise process
        // the child node.
        //
        depthFirst(child);
    }
}
```

Using a search queue, the algorithm follows a different tactic:

Initialize the search queue with the root node.
While the queue contains a node:
Remove the first node from the queue.
If this is the desired node, quit
Add any children from this node to the top of the queue.

Breadth-First

Breadth-first searching goes wide rather than deep. It processes all of the nodes at one level before moving down to the next, as shown in Figure 5-4.

Using a search queue depth-first looks like:

Initialize the search queue with the root node.
While the queue contains a node:
  Remove the first node from the queue
  If this is the desired node, quit
  Add any children from this node to the bottom of the queue.

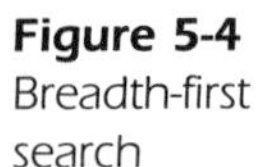

Figure 5-4 Breadth-first search

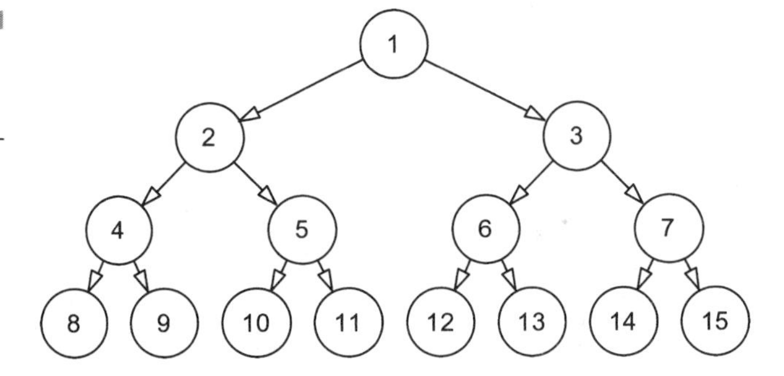

Cheapest-First

This is also known as best-first searching.

If you are searching a state or map space, you may want to take the cheapest route to the goal rather than testing every possible path. At each step in the search, the cheapest-first algorithm takes the branch with the lowest cost. If the path this method finds does not reach the goal, it can back up and start down the next-cheapest branch. The backup could be done in a breadth-first style, going back up to the top next-cheapest node, or depth-first style, visiting the next leaf node.

The cost function is a local test. The cost to the next state is fixed and only the immediate cost of traversing to that next state is used in the decision of which step to take. While this can provide a good solution the local decisions do not guarantee an optimal solution.

A sample cheapest-first search is shown in Figure 5-5. The numbers on the transition arcs are the cost of each move and the numbers in parentheses are the cost for the moves taken up to that state.

The key to the cheapest-first search is in the function that evaluates the costs.

This search makes the most sense if there is more than one way to reach the ultimate goal. The graph should be well connected so the search does not have to backup often. In the worst case cheapest-first degenerates to a depth or breadth-first search.

If the search ends up at a leaf node that is not an acceptable goal the search can back up and continue. Node *a5* represents a minimal backup, while *b5* goes back to the top.

Figure 5-5 Cheapest-first search

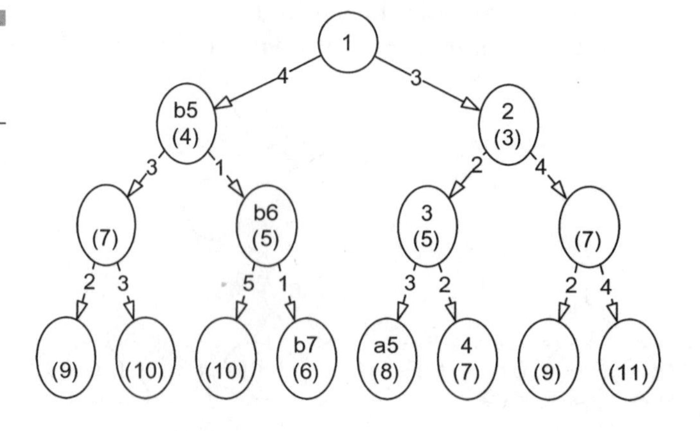

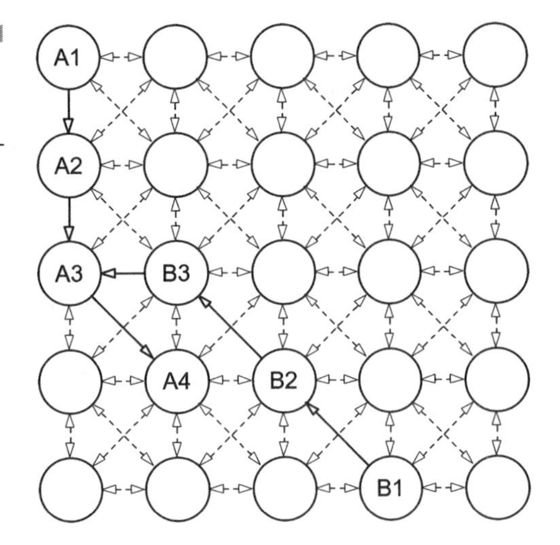

Figure 5-6 Bi-directional search

The solution represented by found at node *4* is not the best solution. It is the best choice based on local decisions, but if the algorithm continues searching it may ultimately find node *b7* with its cost of six.

Bi-directional

A bi-directional search may be the easiest way to find a good path. A bi-directional search using a cheapest-first strategy is shown in Figure 5-6. Note that not all nodes are shown here, only the most relevant ones.

This is an insanely connected graph, perhaps representing paths through an unobstructed environment. This technique also works on graphs with less connectivity or that include obstructions.

The two searches meet when *B3* traverses to the previously visited *A3*. This path still may not be the globally cheapest path, but it is a good path.

A* Search

The A* search, pronounced "A-Star", is used for path finding for robots or in game environments, and for routing traces in electronics CAD systems. It can also be used to find "paths" through state space, traveling from one state to another until it reaches a goal.

A* is a heuristic search. It tries to be clever about not looking at every node in the state graph while still finding an optimum path. Of course, the cheapest-first search also avoids most of its tree but it was too simple minded to generate the best path.

A* is a variation of cheapest-first searching that uses a global cost to determine which node to process next. This avoids the hazard of local minimums that the cheapest-first search falls into. When given the correct cost equations, as discussed later, A* will find the best path. The downside to A* is that it is slower and more complex than the brute-force searches.

This search can be used in both physical space and in abstract state space. Its ability to find routes through complicated and obstructed terrain makes A* valuable for many path-finding applications. Since state space can be treated like physical space, the A* search can also find paths through these abstractions. Whether the nodes in the graph are states or positions in space the search is the same.

At each step in the A* process the cheapest node from a list of open nodes is selected. There are two important differences from the cheapest-first search. One is that in cheapest-first we selected the cheapest next node from the set of nodes reachable by the last processed node. In A* we select the cheapest node from all open nodes, whether or not they are connected to the last processed node.

The other difference is in how the cost for a node is calculated. In cheapest-first we only look at the immediate cost of the next step. In A* the cost of a node includes the cost of all steps leading up to it plus an estimate of how much it may cost to reach the destination from that node. This provides a reality check against the global environment that keeps the search on track for the globally cheapest solution.

The cost estimate to the goal is calculated in different ways for different situations. For maps it may be the Euclidian distance to the goal as the crow flies. For A* to work properly this estimate must be an *under-estimate* of the actual cost to reach the goal.

An over-estimate of the cost to the goal will set the path to wandering like a loose string draped across the map. It is interesting to note, however, that a slightly bloated estimate of the cost to the goal does speed up the search time at the expense of possibly generating a sub-optimal path.

A* is fastest when there is a clear path between the start and the goal. If there are obstacles, it casts its search out to the sides until it finds a way around them. A* is at it slowest if there is no path possible to the goal, since it searches every possible node in the graph to determine that no path exists.

The process of finding a path is illustrated later, in the *Path Finding* section.

We explore two examples of A* in practice. The first example performs A* path finding in a simple grid world. The second example uses A* to navigate through the abstract state space defined by the cuts that can be made in a panel of wood by a panel saw.

First we define the abstract A* search engine.

Code: A* Engine

The state of the A* system is defined by the current active node, a list of open (unexplored) nodes, and a list of closed (explored) nodes.

Each node has a cost composed of two components, the cost accumulated while traveling to this node plus an estimate of the remaining cost to reach the goal. This represents the total distance in state space and the search uses it to build the shortest path between start and goal.
The A* algorithm is:

> Init the open list with the root node.
> While the open list contains nodes:
>   Remove the cheapest node *A* from the open list
>   If this is the desired node, the search is done
>   For all valid nodes *B* connected to node *A*:
>   Assign a cost to *B*
>   Check to see if *B* is on the open or closed list.
>   If so, and this *B* is less efficient, discard it.
>   If not on a list, or this path to *B* is more efficient, more *B* to the open list
>   Add *A* to the closed list.

The structure of the A* search system is illustrated in Figure 5-7. The *aStar* class contains the *m_open* and *m_closed* lists, each of which can contain any number of *aState*-based objects. *aStar* executes the steps of the algorithm described above and *aState* is responsible for finding neighboring nodes and assigning costs.

aState.java

aip.astar.aState *aState* is an abstract class that provides the basic outlines for the state. In addition to specifying the necessary interface methods, *aState* holds the backwards link that is used to remember the path.

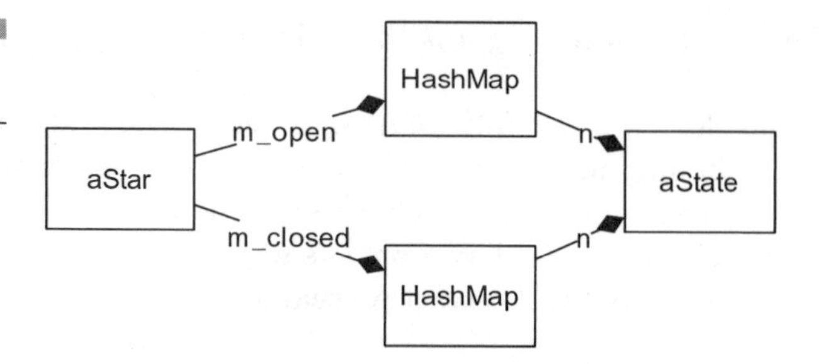

Figure 5-7
A* class diagram

Table 5-1
aState summary

Construction	
	aState ()
Path	
aState	getPrevious ()
void	setPrevious (aState prev)
abstract Iterator	neighbors ()
Scoring	
abstract double	getDistFromStart ()
abstract double	getDistToGoal ()
double	getCost ()
State	
abstract boolean	done ()
abstract Object	getKey ()
abstract boolean	equals (aState state)

A summary of *aState* is given in Table 5-1.

aState() The generic *aState* does not need anything for its constructor. However, application specific states may need access to a common context, such as a map, as well as local information to define the state.

getPrevious()
setPrevious() Each state maintains a backwards link to the previous state.

neighbors() Each state knows about its neighbors and must be able to provide an iterator that contains all of its legal neighbors.

getDistFromStart() When a new state is created in the *neighbors()* method, the costs accumulated by the path so far are stored in the new state. This method reports on this cost.

getDistToGoal() Calculate and return an estimate of the remaining distance to the goal. An underestimate of the true distance creates an optimum path while a slight overestimate speeds up the search.

getCost() *getCost()* returns *getDistFromStart()* + *getDistToGoal()*.

done() The state must be able to tell if it is the goal state or not. If it is the goal state, it returns *true*.

getKey() States are stored in hash tables to facilitate lookup in the open and closed tables. The *getKey()* method returns the application-specific key used to identify a state.

equals(aState state) Each state must also be able to determine if it represents the same condition as another specified state.

aStar.java

aip.astar.aStar *aStar* implements the A* algorithm. The application-specific information is contained or referenced through the *aState* child classes, so *aStar* itself is completely generic.

A summary of *aStar* is given in Table 5-2.

aStar(aState start) The start, or root, node is the starting point of the search and must be specified in the *aStar* constructor. The goal is an attribute of the problem space and is implicit in, or attached to, the states.

search() This executes the A* algorithm, spawning states and selecting them until a path is found to the goal or it gives up in failure. If *search()* succeeds it returns an *ArrayList* containing the states in the path, including the start and the goal. If it fails it returns *null*.

getStepCount() In order to tune the parameters of the A* search, it helps to know how much work it is doing. The *getStepCount()* method returns the number of states *search()* has touched before finding its solution.

Table 5-2

aStar summary

Construction
```
          aStar (aState start)
```
A* Search
```
ArrayList search ()
      int getStepCount ()
```

Code: A* Path Finding

While the generic A* engine is interesting the real fun begins when you can watch it in operation. This example applies *aStar* to the task of finding a path across a map grid. We also discuss what makes a good path and how to adjust the cost evaluations to nudge the algorithm into creating one.

The map is shown in Figure 5-8. The cost of moving between two squares in this map is one. Moving diagonally to a square has a cost of $\sqrt{2}$. This is only one way a map may be defined and scored, and it is not necessarily the most efficient way.

On a large map there may be an unreasonable number of grid points, making the path search unwieldy. One optimization is to define the map as a quadtree, providing views of the space at different scales. The initial path could be found at a high level and then sub-paths could be found to fill in this meta-path.

The map may not be a grid at all. It could be a continuous space partitioned by triangular patches. Or it could be a set of pre-calculated waypoints that follow natural map features. A number of excellent articles discussing map representations and map-based optimizations are found in DeLoura (2000) and Rabin (2002). Searches with moving targets and real-time A* techniques are discussed in Weiss (1999). The A* path finder is also a good candidate for bi-directional searching.

We, however, describe a basic search on the grid-based map from Figure 5-8.

The points in the path lie on the intersections of the grid lines. The bottom-left of the map is point (0,0) and the top-right is (15,15). The starting point is (1,1) and the goal is (14,14).

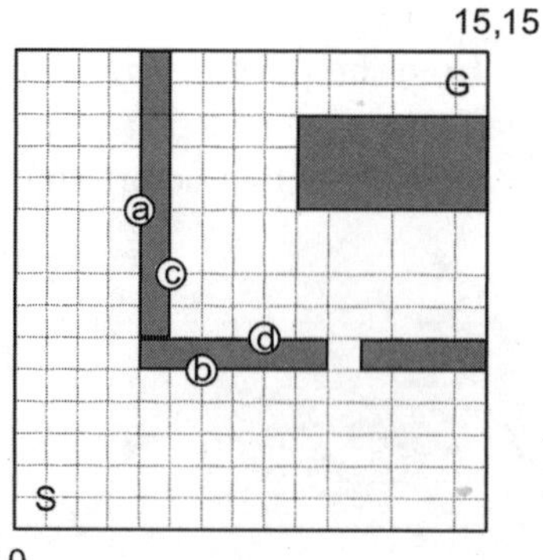

Figure 5-8
A* map

The *Rectangle* class that represents the barriers is inclusive on the left and bottom. Points on the left or bottom edges of the barrier (points *a* and *b*) are excluded from the path. Points on the right or top edges of a barrier (*b* and *d*) are not in the barrier and may be used in the path.

This actually leads to a flaw in the barriers. Since the walls are one step wide, and points on half of the wall's edges are valid positions, the step from (10,5) to (9,6) is allowed though it clearly passes through the corner of the wall. A small change to the *Rectangle* class to reject all points on the edges would fix this problem, but make the current gap in the wall impassible.

Using the cost function described earlier, A* finds the path shown in Figure 5-9(a). This path appears to be a bit wiggly. It does not conform to our idea of an efficient path. By penalizing turns in the path with an added 0.5 cost, the path straightens out to the one shown in Figure 5-9(b).

The scoring defines the type of path the search finds. If you are working with circuit layouts, you may want only 45° turns to avoid sharp corners. To lay traces near each other, the cost at each step may be reduced if it is near an existing trace and increased everywhere else.

In games, the travel costs may vary based on what environment is found there; whether it is a clear path, a rough and rocky area, brambles, etc.

Path 5-9(b) is found after processing 173 states. By doubling the value returned by *getDistToGoal()* the processing is cut by a third, finding the same path while analyzing only 54 states. This difference is illustrated in Figure 5-10. This shows how the search pulls into a tighter group between the start and the goal as the distance cost is increased.

There is a negative side effect to this efficiency, however. If we create an opening in the vertical wall, the double-cost search misses it. The unit cost search, however, successfully finds this new, shorter, path (Figure 5-11).

Figure 5-9
A* paths. Simple distance cost vs. additional turn penalty

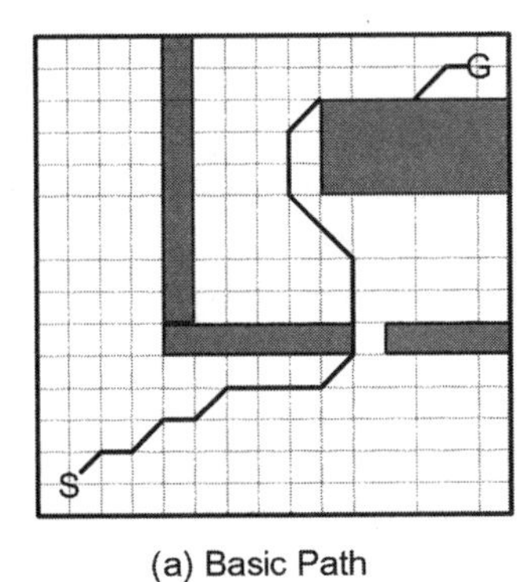

(a) Basic Path

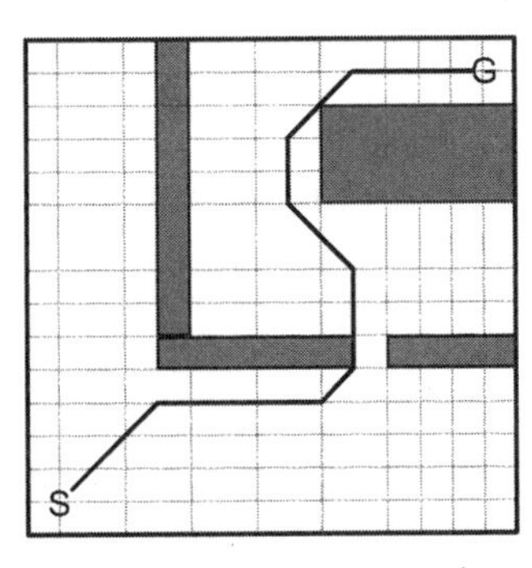

(b) Added turn penalty

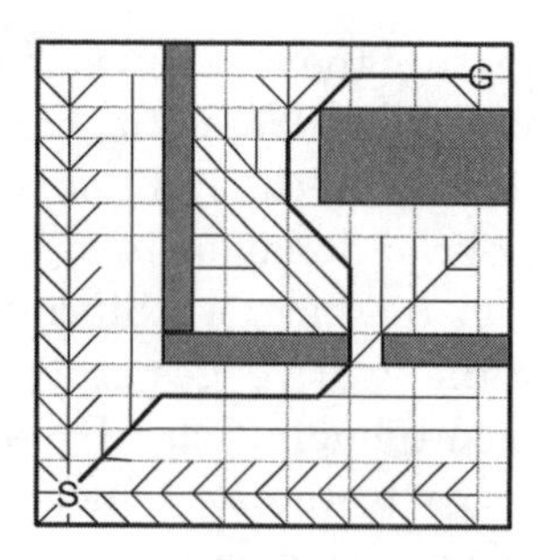

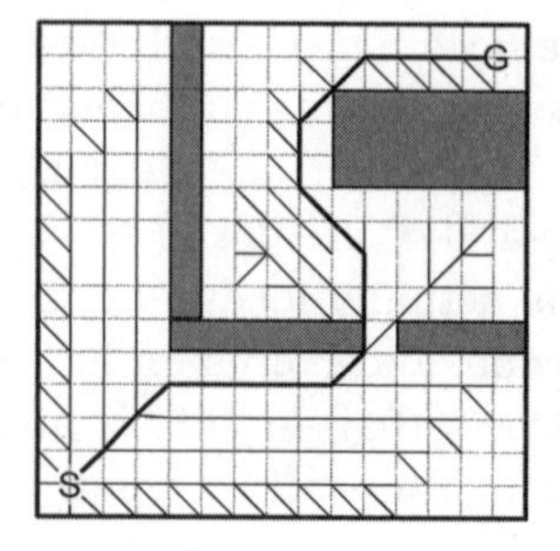

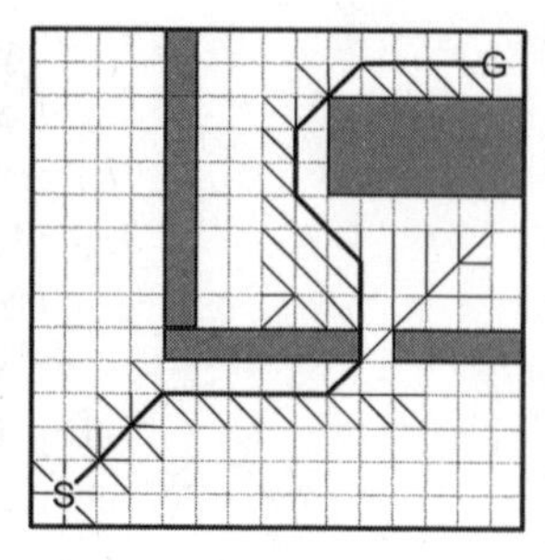

Figure 5-10
A* processing

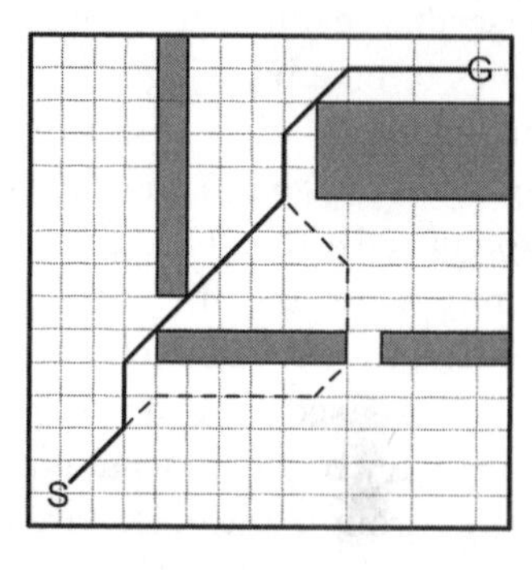

Figure 5-11
New door

PathState.java

aip.app.pathfinder.PathState The *PathState* class adds map knowledge to the generic *aState*. Both the size of the map and the list of barriers are stored as static attributes of *PathState*. An alternative would be for *PathState* to reference an external map management system.

A summary of *PathState* is given in Table 5-3.

There are only a few methods in *PathState* that are not in *aState* (Table 5-1). We discuss a few selected methods here.

PathState(Rectangle world, Point origin, Point goal) The constructor for *PathState* sets up the world and the two endpoints in the path.

addBarrier(Rectangle barrier) The *addBarrier()* method drops rectangular dead zones into the world map. This information is stored in *PathState* as static data so it is not duplicated across the states created during processing.

getKey() The key for a *PathState* is the *X*, *Y* coordinate of that state's position on the grid. Any state with the same *X*, *Y* position is considered to be equal to this state.

Table 5-3

PathState summary

```
Construction
             PathState (Rectangle world, Point origin, Point goal)
        void addBarrier (Rectangle barrier)

Path
    Iterator neighbors ()

Scoring
      double getDistFromStart ()
      double getDistToGoal ()

State
     boolean done ()
      Object getKey ()
     boolean equals (aState state)

Access
       Point getPos ()
         int getBarrierCount ()
   Rectangle getBarrier (int idx)
      String toString ()
```

getPos()
getBarrierCount()
getBarrier(int idx) The access methods provide a way to retrieve information from the *PathState*.

PathFinder.java

aip.app.pathfinder.PathFinder The *PathFinder* application is a command line program. It creates the root state and the barriers, and then initializes the *aStar* object *star* with the root state. The search itself is a single call, *star.search()*.

After the search it prints the resulting path coordinates to the console.

Code: A* Panel Nesting

This is an abstract problem where the nodes represent steps along a process and not positions on a map.

The classic state-based A* problem is the fifteen puzzle. This puzzle is a 4×4 frame with 15 sliding squares in it plus a blank space. The solved position puts the "1" square in the top left and counts the tiles down to 15. The blank goes in the bottom right corner. See an example of this puzzle in Ballard (1997).

Each state in the tree represents the state of the puzzle after a single piece is moved. The cost from the start is the number of moves made so far.

The cost to the goal can be the count of how many squares are out of place, or perhaps the Manhattan distance of each square to its proper position.

The example we use is a simplified form of a common industrial problem. This problem can take different forms, ranging from cutting pieces from a stick of material to the optimum organization of boxes in the hold of a ship.

This example organizes rectangular parts on a larger rectangle of material.

The goal is to take a sheet of material, called the panel, and a list of rectangular parts and cut the parts out of the panel with a panel saw. A panel saw can only make cuts all the way across the material. After a cut you are left with two pieces, one that has one or more parts on it and another that is any remaining panel.

Say you have a 20×10 panel and you want to place three parts on it. The parts are each 6×5. These parts can be removed with the four cuts shown in Figure 5-12. Cut (1) separates the panel from the parts. Cuts (2), (3), and (4) separate the parts. The result is shown in Figure 5-13. Once the three parts are removed we are left with a 5×20 panel and a 2×5 scrap.

In order to keep the example simple we make some simplifying assumptions. The first is that we only track one sheet of material, so all scrap pieces are discarded. In a real application parts could be cut out of the scrap.

As a result of this assumption, it is implied that parts do not have holes in them that could have smaller parts placed in them.

With these assumptions in place we can define our states and the two cost measures. Each state consists of a panel and a single cut on the panel. It may be safely assumed that no two states in a sequence are the same.

Figure 5-12
Panel cuts

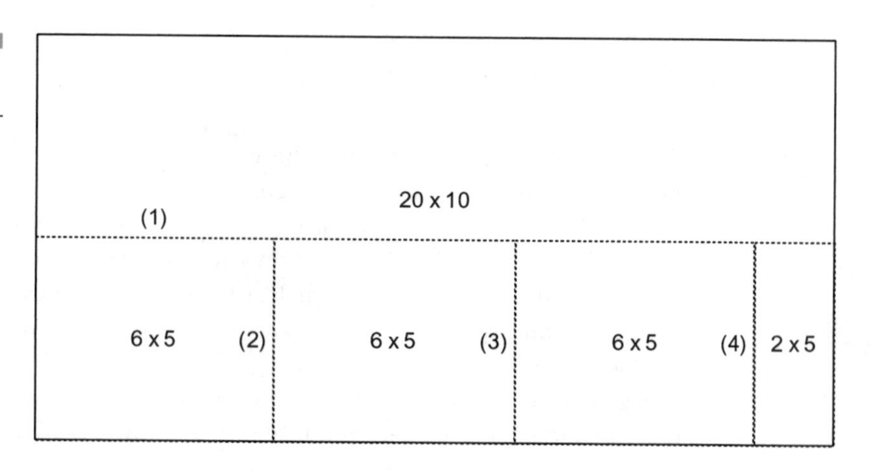

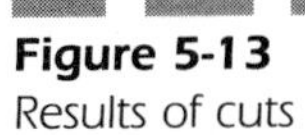

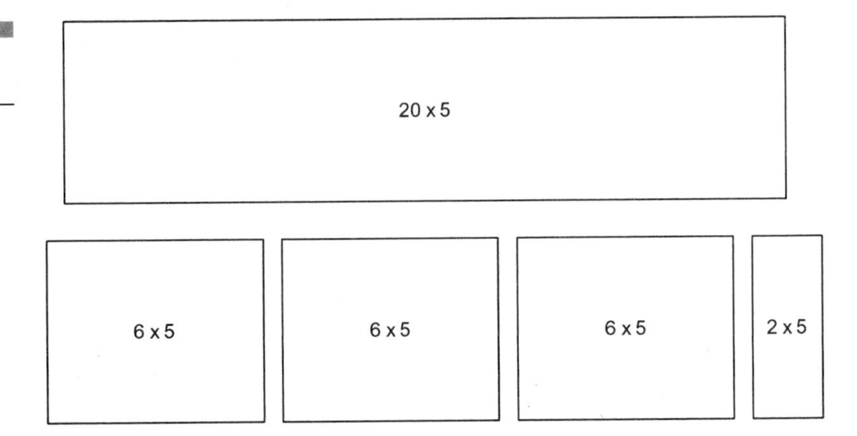

Figure 5-13
Results of cuts

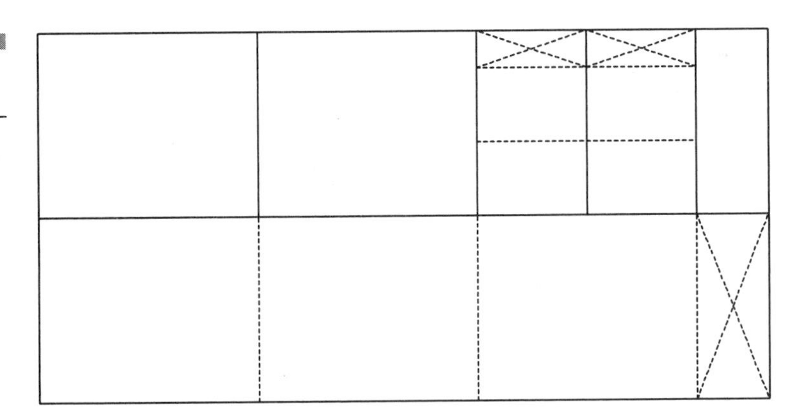

Figure 5-14
Panel nest results

The cost from the start is the area of all scrap discarded up to this point. The cost to the goal is the area of the remaining uncut panel.

An additional requirement is that all of the parts must fit on one panel. The nest may not spill over onto a second sheet, though a useful version of this program would use as many sheets as needed to cut all of the parts.

Given our 20×10 panel we try to fit two sizes of part on it; five each of a 6×5 part and four each of a 3×2 part. Because of imaginary grain restrictions, we do not rotate these parts during the nest. In other applications it is possible to rotate the parts to find better nests.

Running the *PanelNest* application gives the result shown in Figure 5-14. If you look at the actual program output you may notice that it only lists the major cuts. It does not bother to separate the individual parts from a strip.

Table 5-4

PanelState summary

Construction	
	PanelState (Rectangle2D panel)
void	addPart (double width, double height, int num)
State Path	
Iterator	neighbors ()
double	getDistFromStart ()
double	getDistToGoal ()
State	
boolean	done ()
Object	getKey ()
boolean	equals (aState state)
Access	
Rectangle2D	getPanel ()
int	getPartCount ()
Rectangle2D	getPart (int num)
int	getPartQty (int num)
Line2D	getCutLine ()
int	getCutPartNum ()
int	getCutPartQty ()
Rectangle2D	getCutPart ()
boolean	getCutHorz ()
String	toString ()

PanelState.java

aip.app.pathfinder.PanelState The *PanelState* is more complex than the *PathState* was since it has to remember more information. The state must remember its costs, the remaining panel to be cut, the cut the state is proposing, and the remaining quantity of each part. The only thing common across all states is the list of part sizes.

Like *PathState*, *PanelState* has essentially the same interface as *aState*. The meanings of *neighbors()*, *getDistFromStart()*, and *getDistFromGoal()* are all specific to the panel nest application and a new set of access methods has been sizes.

A summary of *PanelState* is given in Table 5-4.

Two-Player Games

In the A* search we optimized a path through state space. As the search progressed the cost followed a steady progression. The historical cost was continually growing and the predicted cost was shrinking.

In a two-player situation, such as any turn-taking game, the evaluation of the cost of a move reverses at each step. Player A, our point-of-view player, wants to maximize our score while player B, the opponent, wants to maximize *their* score (which minimizes player A's score).

For simple games, such as tic-tac-toe, it is possible to generate a tree of all possible moves up to the end of the game. The leaf nodes on this tree would be marked win, lose, or draw. The computer could then choose moves that always led to a win or, at worst, draw, because it has perfect knowledge of all possible game states.

This tree of all possible states is the game tree. The current state of the game at the time we are making a decision is the root node of the tree. Each subsequent state is the result of a move in the game. The player who moves first is our player, A, who wants the maximum score, so that is the *max* player. The other player is the *min* player. Each pair of moves, *max* then *min*, is one ply.

Most games have far too many possible states to allow a complete search. Chess, which is a popular example of rational games, has roughly 35 possible moves at each ply. After just a few plies this creates billions of possible game states.

Instead of using an exhaustive search most game searches look a limited number of moves ahead and choose the most promising path.

Min–Max Search

Min–max is a two-player game search that finds the most promising path in a game while looking ahead a limited number of turns. This search is named after the conflicting goals of the two players, where one wants to maximize the score and the other seeks to minimize it. The score itself is a comparison of what is good for the *max* player minus what is good for the *min* player.

The skeleton of min–max is a simple depth-first search that is limited to a fixed depth.

There are two ways each node in the tree can be scored. If the search has reached its maximum depth or a dead-end state, the value of that node is the static score of that game state.

Nodes that are not leaves in the search are scored differently depending on whose turn they represent. Nodes for *max* player's turns are max nodes, and they take on the value of their highest scoring sub-node. Min nodes take on the value of their lowest scoring sub-node.

The development of a good scoring system for a game is an art in itself. While simple games like tic-tac-toe are easy to evaluate, complicated games like chess or, even worse, Go, can have very complicated and subtle conditions of goodness and badness.

Because of its simplicity, tic-tac-toe is used for our illustrated examples.

A tic-tac-toe board consists of rows, columns, and diagonals, known collectively as spans. A player wins by filling a span with their mark, either *X*'s or *O*'s.

The static score for any given tic-tac-toe arrangement consists of the spans the max player, *X*, can still use to win minus the number of spans that the min player, *O*, can still use to win. One *X* in an empty span is good, however, two *X*s in an empty span are even better, so spans with two of the same mark get valued higher:

$$\text{score} = (3X_2 + X_1) - (3O_2 + O_1) \tag{5-1}$$

where X_2 is the number of spans where there are two *X*s plus a blank on a span and X_1 is the number of spans with one *X* plus two blanks. The same is true for the *O* scores.

Hence, the board arrangement in Figure 5-15 has a score of:

$$\text{score} = (3*1 + 3) - (3*0 + 2)$$
$$\text{score} = 4$$

The execution of the min–max algorithm is easiest to describe as a code outline:

```
Move findMove(GameState root, int lookahead)
{
    double best_score = Double.NEGATIVE_INFINITY;
    Move best_move = null;

    Iterator move_iter = root.generateMoves();
    while (move_iter.hasNext())
    {
        Move this_move = (Move)move_iter.next();
        //
        // Root is a max node
        //
        double score = this_move.minmax(lookahead, true);
        if (score > best_score)
```

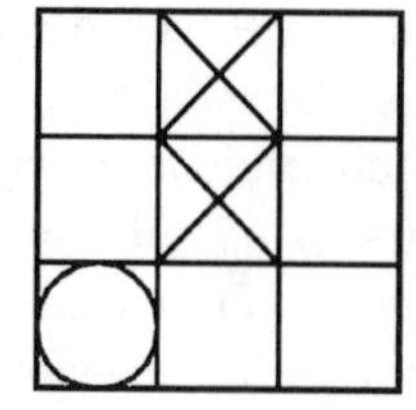

Figure 5-15
Tic-tac-toe scoring

```
            {
                best_move = this_move;
                best_score = score;
            }
        }
        return best_move;
    }

    double minmax(
        int depth,
        boolean maxnode )
    {
        if (depth == 0)
        { return staticScore(); }

        double score = Double.NEGATIVE_INFINITY;

        Iterator move_iter = generateMoves();
        while (move_iter.hasNext())
        {
            Move next_move = (Move)move_iter.next();

            if (maxnode)
            { score = Math.max(score, next_move.minmax(depth-1,
            !maxnode)); }
            else // minnode
            { score = Math.min(score, next_move.minmax(depth-1,
            !maxnode)); }
        }

        return score;
    }
```

There are two pieces to the code. The first part, *findMove()*, selects the branch that contains the highest valued move. The *minmax()* call recursively scores each branch.

Looking at the first ply of the first move in tic-tac-toe, we get a graph like that shown in Figure 5-16. Note that symmetrical positions are ignored for clarity.

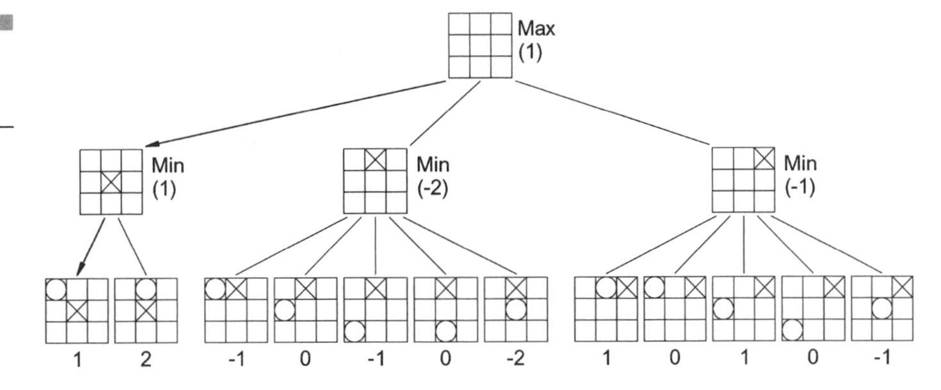

Figure 5-16 Min–max tree to one ply

Using a depth-first scan we move down the left edge to the first leaf on the left, which scores as a 1. Its neighbor scores a 2. Since the parent of these is a *min* node it uses the smaller score of 1. The other mid-level *min* nodes take on the values of –2 and –1, respectively. These represent the best moves the *min* player can make for the moves at this level.

The *max* player, the computer, chooses the largest of these *min* scores. This is the *X* in the center move, since no matter what the *min* player does we get a better score than can be expected from the other moves.

Alpha–Beta Pruning

In the min–max example earlier, the best path was chosen using a limited look ahead. But what if there is some terrible turn of events beyond the horizon that might turn the tables in favor of the opponent? The farther ahead the game looks the better the decision it can make. Unfortunately, the set of game states explodes with even a limited look ahead.

Alpha–beta pruning provides a safe way to give the computer a deeper look ahead at the expense of the width of its search. It also guarantees that the branches it prunes from the game tree will not affect the outcome of the search. They are provably dead branches and removing them speeds up the search without negative side effects.

The alpha and the beta of this algorithm, α and β for the Greeks among us, track the results at each level of the search. α ratchets up to match the highest score found in a *max* node and β ratchets down to match the lowest score found in a *min* node. If at any point α exceeds β the rest of the sub-nodes and their descendents can be ignored. We have already found the best strategy for this branch and nothing else can improve it.

The algorithm for alpha–beta follows the same basic plan as min–max, with some changes in the score accumulation. Note that the code presented here is not the most efficient way to implement alpha–beta, but it illustrates the logic of the problem. It is possible to remove the min/max by restructuring things so that the core call looks like:

```
score = -next_move.alphabeta(depth-1, -beta, -alpha));
```

But since alpha–beta is confusing enough as it is, this variation is left to the reader. You can track down Bruce Moreland's description of it at `www.seanet.com/~brucemo/topics/alphabeta.htm`.

The root node in the alpha–beta search is called with infinite values for the alpha and beta limits:

```
double score = this_move.alphabeta(lookahead, true,
                      Math.NEGATIVE_INFINITY, // initial alpha
                      Math.POSITIVE_INFINITY);// initial beta
```

The scoring, with appropriate pruning, is outlined here:

```
double alphabeta(
   int depth,
   boolean maxnode,
   double alpha,
   double beta )
{
   if (depth == 0)
   { return staticScore(); }

   double score;
   if (maxnode)
   { score = Double.NEGATIVE_INFINITY; }
   else
   { score = Double.POSITIVE_INFINITY; }

   Iterator move_iter = generateMoves();
   while (move_iter.hasNext())
   {
      Move next_move = (Move)move_iter.next();

      double this_score = next_move.alphabeta(depth-1, !maxnode,
      alpha, beta);
      //
      // Shift the score and alpha/beta values the correct
         direction
      //
      if (maxnode)
      {
         score = Math.max(score, this_score);
         alpha = Math.max(alpha, score);
      }
      else
      {
         score = Math.min(score, this_score);
         beta = Math.min(beta, this_score);
      }
      //
      // Prune?
      //
      if (MathConst.isLesser(beta, alpha))
      { break; }

      if (alpha >= beta)
      { break; }
   }
   return score;
}
```

Figure 5-17
Alpha–beta, moves to center X

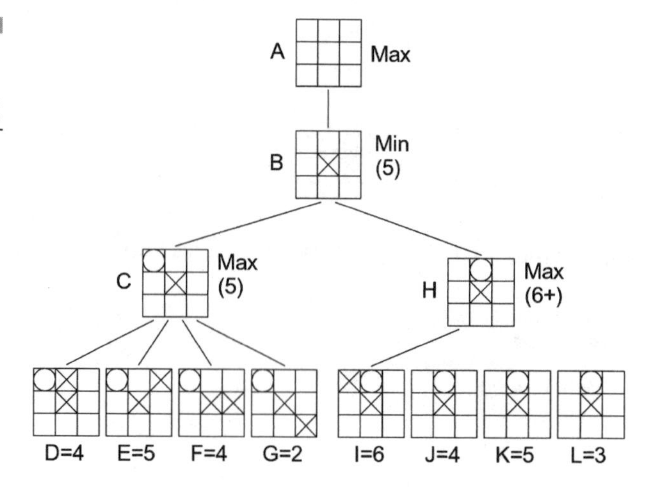

The *findMove()* half of the search, not illustrated here, calls *alphabeta()* and is roughly same as in the min–max search. The branch scoring is like min–max with the addition of the alpha and beta thresholds and pruning.

This can be hard to visualize from just the algorithm, so let us take a tour through a tic-tac-toe game that has a look ahead of three moves.

There are three unique first moves available in the game: *X* in the center, edge, or corner. Let us watch the alpha–beta process the branch where *X* is in the center. Figure 5-17 shows this branch of the game tree.

The root node *A* calls *alphabeta()* on *B* with $\alpha = -\infty$ and $\beta = \infty$. *B* descends to *C*. *D* returns a static score of 4, so at the *C* level our α now equals 4. *E* scores 5 putting α at 5. *F* and *G* score lower than 5, so *C* returns and takes on the max score of 5.

After processing *C* the parent node *B* moves its β down to 5 and calls *H* with $\alpha = -\infty$ and $\beta = 5$.

In *H* we process sub-node *I* with its score of 6. This puts α at 6 which is greater than the β of 5. No matter what other score we get *min* node *B* will not pick *H*. *C* has a lower score and will always have a lower score. We are done and can ignore nodes *J*, *K*, and *L*.

B returns its minimum score of five up to root node *A*. At *A*, α is now 5 and β is still ∞. The next branch is shown in Figure 5-18.

The previous branch returned a score of 5 to node *A* and set $\alpha = 5$ and $\beta = \infty$. Node *A* now calls *b* with $\alpha = 5$ and $\beta = \infty$, and *b* passes this down to *c*.

Node *c* scans its nodes *d* through *j* and finds the maximum score of 4. This is returned up to *b*.

Since *b* is a *min* node, it moves its β down to 4. Note that $\alpha = 5$ and $\beta = 4$, making α greater than β. It does not matter what the sub-trees

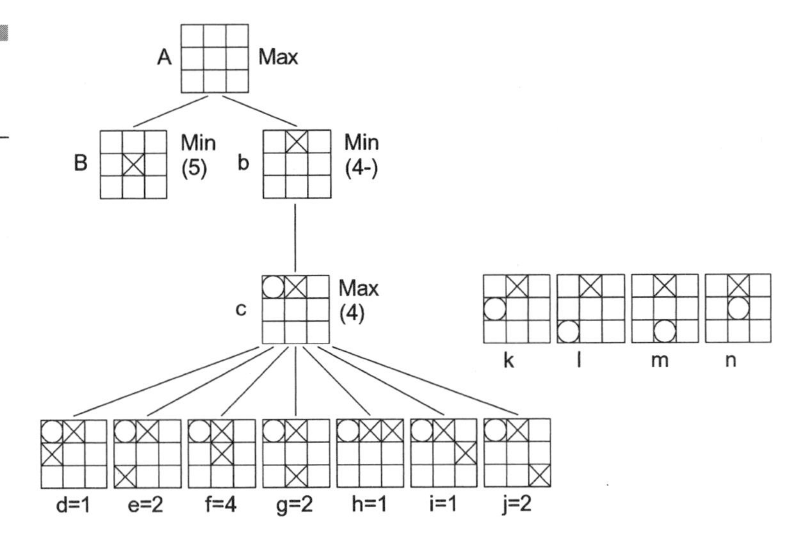

Figure 5-18 Alpha–beta, moves to edge X

headed by nodes *k* through *n* score at, the *min* node *b* will not pick anything with a larger value than *c* node's score of 4, which means that *max* node *A* will not pick anything in this branch at all since the previous branch headed by *B* returned a better score of 5. Whew!

The remaining branch with *X* in the corner offers more of the same. The best move ultimately turns out to be *X* in the center with its score of 5.

Code: Alpha–Beta Engine

The *AlphaBeta* class does its work using application-specific implementations of the interfaces *abGame* and *abMove*. Though the alpha–beta search does not care about the details of the application, it does operate in the world of two-player min–max games; so the terminology is in terms of games, scores, and moves.

AlphaBeta.java

aip.alphabeta.AlphaBeta The interface to the *AlphaBeta* search is the single method, *findMove()*.

findMove (abGame game, int lookahead) *findMove()* takes the game in progress and looks ahead the specified number of moves to try and find the best next move to make.

There is one significant difference in how the alpha–beta algorithm is implemented versus its previous description. Instead of holding the entire game state in *abMove*, there is a stack of moves in *abGame* that keeps a history of past moves. These moves can then be unmade.

```
game.pushMove(this_move, !maxnode);
double this_score = alphabeta(game, depth-1, !maxnode, alpha,
beta);
game.popMove();
```

abGame.java

aip.alphabeta.abGame All of the work in the search is performed by an application-specific implementation of the *abGame* interface. This interface manages moves and performs balance-of-power scoring.

A summary of *abGame* is given in Table 5-5.

scoreBalance() The value returned by *scoreBalance()* indicates the balance of power in the game. This is not necessarily the piece count. Evaluating who has the upper hand can be a subtle and difficult task.

This method is the heart and soul of an alpha–beta game. If the scoring is bad the move solutions will be bad.

scoreMove(abMove move, boolean max_node) In addition to scoring the balance of power, it sometimes helps to score the tactical value of a move itself.

generateMoves(boolean max_node) To search the game tree, you must be able to create the game tree.

pushMove(abMove move, boolean max_node) During game tree evaluation, it is necessary to change the game state to see the effect of a move. In this implementation there is only one game state. The moves modify this state.

When a move is pushed, the game state is modified. A record of the move is made so it can be undone with *popMove()*.

popMove() Undo a previously pushed move.

Table 5-5

abGame summary

```
Public Methods
  double  scoreBalance ()
  double  scoreMove (abMove move, boolean max_node)
Iterator  generateMoves (boolean max_node)
    void  pushMove (abMove move, boolean max_node)
    void  popMove ()
```

abMove.java

aip.alphabeta.abMove This is an empty interface.

Code: Amoeba Game

While we could write an easy game like tic-tac-toe, or a slightly more involved game like Connect-4, we instead invent a game, Amoeba.

Amoeba illustrates the use of alpha–beta and it also highlights some of its weaknesses.

Amoeba is played on a hexagonal grid, like that shown in Figure 5-19. Though the grid could be any size, this one is a manageable 4×6 cells.

Players take turns placing their pieces in the grid. At the end the player with the most pieces wins.

If a piece is sufficiently outnumbered by its opponent's pieces, that piece is "killed" and removed from the board. Each cell in the grid has up to six neighbors. To determine if a piece is killed, count the number of friendly neighbors and the number of opponent neighbors. If the opponents outnumber the friendlies by two or more, that piece is removed from the board.

Simple, right?

The difficulty lies in determining what is a good move and what is a bad move.

The number of pieces, their connections to other pieces, and even their position on the board all affect the balance of power in this game.

Let us take a tour of the *Amoeba* program and see what it does. The *Amoeba* class diagram is shown in Figure 5-20.

Since Java is not the speediest of languages, and this *AlphaBeta* implementation is not the most efficient, our look ahead is fairly shallow.

Figure 5-19
Amoeba grid

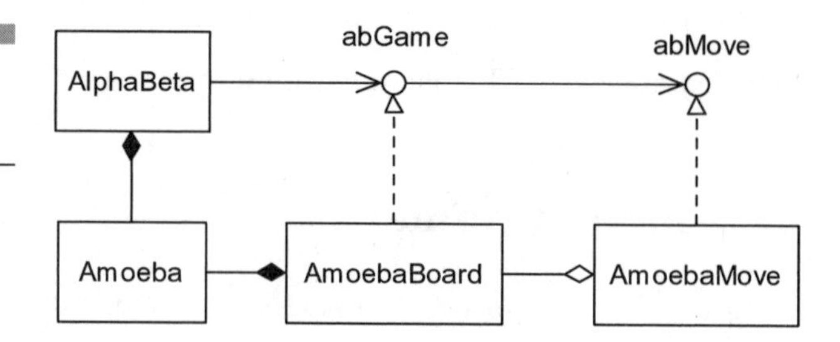

Figure 5-20 Amoeba class diagram

You could adjust the look ahead dynamically, going deeper as the number of possible moves decreases. Or you could keep track of the time taken by each move and try to operate in constant time, regardless of the search depth.

You could have different scoring or look ahead strategies for the start of the game, the mid-game, and the end-game. Of course, the scoring can be adjusted and tweaked endlessly.

These variations to improve the computer's play are left as an exercise for the reader.

AmoebaBoard.java

aip.app.amoeba.Amoeba All of the game logic is held in the *AmoebaBoard* class, so it has the most intricate interface. It knows about valid moves, how to perform those moves and evaluate their side effects, and it knows the value of moves and board configurations. It also draws the board to the display.

A summary of *AmoebaBoard* is found in Table 5-6.

AmoebaBoard(int width, int height) An instance of *AmoebaBoard* is good for one game and it must be started with the board's dimensions.

paint(Graphics2D g2) This paints the entire board to the graphics display. There is no compelling reason to use *Graphics2D* instead of *Graphics* in this application, though in other situations *Graphics2D* provides a wider range of drawing tools.

validMovePos(Point pt, boolean red_move)
validMove(int x, int y, boolean red_move) Determine if the move is valid for the specified player. *validMovePos()* checks the move at the mouse position and *validMove()* is passed to the actual cell coordinates.

A valid move is one that is not already overpowered by an opponent's pieces.

Table 5-6
AmoebaBoard summary

```
Construction
            AmoebaBoard (int width, int height)

Visuals
        void  paint (Graphics2D g2)

Moves
     boolean  validMovePos (Point pt, boolean red_move)
     boolean  validMove (int x, int y, boolean red_move)
        void  makeMovePos (Point pt, boolean red_move)
        void  makeMove (int x, int y, boolean red_move)
        void  showMovePos (Graphics2D g2, Point pt)
    Iterator  generateMoves (boolean red_move)
        void  pushMove (abMove move, boolean red_move)
        void  popMove ()

Scoring
      double  scoreMove (abMove move, boolean red_move)
      double  scoreBalance ()
         int  scoreGreen ()
         int  scoreRed ()
```

makeMovePos(Point pt, boolean red_move)
makeMove(int x, int y, boolean red_move) Place a piece on the board of the specified color. *makeMovePos()* drops the piece at the mouse's position while *makeMove()* receives the cell coordinates.

After the move is made the internal *evaluate()* method is called. This scans the entire board and removes pieces that have too many enemy neighbors. They have been digested by the all-powerful amoeba!

generateMoves(boolean red_move) Return a list of all cells that pass the *validMove()* test for this player.

pushMove(abMove move, boolean red_move) Essentially the same as *makeMove()* except that it records the move on the move stack. *evaluate()* also records any cell deaths resulting from the push on the stack. The moves pushed onto the stack as a direct consequence of *pushMove()* are called key moves. The side effects of the move caused by *evaluate()* are not key moves.

popMove() Pop undoes moves from the move stack until it pops a key move. This un-does all of the changes made by *evaluate()* and *pushMove()*.

scoreMove(abMove move, boolean red_move) Each move has a tactical value independent of the effect it has on existing pieces and *scoreMove()*

tries to determine that value. In *Amoeba* the computer does not like to move on the edges and, in fact, prefers to move near the center of the board. It tries to cluster its pieces but not clump them into a tight knot. Finally, it really likes to take the opponent's pieces.

scoreBalance() While the game score is a simple count of red pieces versus green pieces the *scoreBalance()* also looks at the connectivity of the pieces. Lone pieces are weak and vulnerable while groups of pieces are strong. Negative values favor green and positive values mean red is doing better.

The evaluation of the game board is one of the hardest parts of game AI. What is good? What is bad? The evaluation used here is a hand-built heuristic. It is also possible to apply neural net or other learning methods to discover the value of different board configurations.

scoreGreen()
scoreRed() These simply count the number of green or red pieces on the board.

CHAPTER 6

Searching State Space

This chapter extends our idea of state space from the discrete representation used in Chapter 5 to an infinitely variable continuous space. Several methods for dividing state space into manageable forms are discussed.

Two techniques for exploring continuous state space are explored. Reinforcement learning uses occasional feedback to find its way to a good solution, while genetic algorithms use a fitness measure to breed the best solution.

What is State Space?

The states in the finite state machines and Markov systems from Chapter 4 were explicit, defined as nodes in a graph. In Chapter 5 the states were still nodes in a graph, but they represented something deeper, such as an arrangement of pieces in a game or a position on a map. In general, a state represents the current settings of a system. For a game it is the position of all the pieces. For a process it may be a history of all decisions made to that point. For a mechanical system it could be the position and velocity of all of its components. For a software system it could be the value of all of the parameters defining its operation. State is many things to many people.

State space is the collection of all of the possible states for a system. For a game, that would be every possible arrangement of its pieces. For control system with n parameters it is the n-dimensional space defined by those parameters.

Chapter 5 operated in a fairly limited, discrete state space. This chapter starts to move into the more complicated territory of continuous state spaces where the states are not necessarily countable, but are instead defined by infinitely variable parameters.

The easiest way to deal with continuous space is to partition it into discrete cells, transforming it into a discrete state space. This is not always appropriate, however, so the developer is sometime stuck with the more difficult job of working directly in continuous space.

While much of this chapter works in a partitioned state space, we also approach the problem of working in continuous space.

Reinforcement Learning

The previous search methods, such as A*, were finding their way through state space by evaluating each state with a cost, score, or value function.

The state with the best evaluation was used as the next state. This process proceeded, in various ways for various searches, until the goal was reached.

The presumption of these searches is that there is a known cost function, a way to define the relative merits of each and every state. But what if there is no way to know the value of a state in advance? What if there is no explicit model of the state space and our processes interaction in it?

In the case where we do not have a pre-defined value function we can instead learn one and Reinforcement Learning (RL) is one way to do this.

There are two prerequisites a problem must meet to be solved by RL. The problem must be cyclic, because RL improves its performance over time. A repeating cycle is necessary for the behavior to converge onto a correct solution. And, while RL does not require value, or reward, feedback at each step of the process it must *occasionally* receive reward feedback; even if this feedback only occurs at the end of each iteration, indicating success or failure.

The occasional rewards are used by RL to tune its path from the source state to the goal. The first time the system may find the goal through sheer blind luck. The next time, using what it learned from the first success, it will be more efficient, and so on.

RL may be applied to many types of problems, like A* in Chapter 5. Because it is a simple way to build up our intuition of the process, we will develop the RL algorithm with respect to a map.

Core Concepts: Temporal Difference Learning

Let us look at the concepts of RL with respect to a variation of the map used during A* path finding (Figure 6-1).

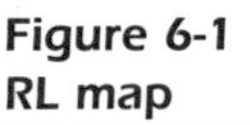

Figure 6-1
RL map

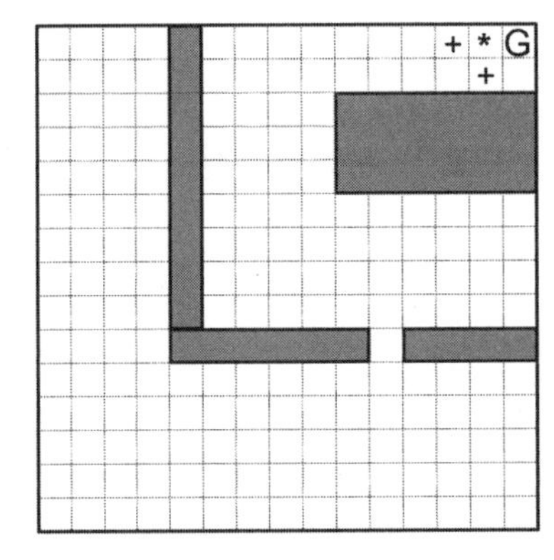

RL works within a set of states. In this example, each square of the map is a state. For each state there will be a set of legal actions, in this case a move North, West, South, or East. At the edges of the map, or at walls, some of the moves will be marked as illegal, limiting the action choices.

The data to be stored is the number of states times the number of actions per state. As you can tell, RL is very sensitive to combinatorial explosion. The data size grows rapidly with additional states or actions, making it necessary to find ways to store only the relevant states and actions.

The actions and states create this mapping:

$$a_s \rightarrow s' \quad \textbf{6-1}$$

where the action a_s taken from one state leads to a new state, s'.

This definition follows the rules of a Markov decision process (MDP). As shown so far, each action has an equal probability of being taken. Starting at a random start state and taking a random walk on the map, it could take a very long time to reach the goal. It may not, in fact, ever get there, though we could nudge it along by discouraging the use of the same a_s as it wanders across the state space.

A learning process is used to adapt the action probabilities so that after the first walk through the map, a second walk will be more efficient. Let us add one more datum to the map, a value for each state V_s. All states are initialized to zero value or, perhaps, a small random value. The goal state gets some larger value such as 10. Upon entering a state the system receives a reward r_s, also called a return. The goal of the system is to maximize the rewards it receives over time.

Once the system steps from a neighboring state $*$ onto the goal G, it receives a reward r (Figure 6-1). It can use this reward to remember that the previous state leads to a reward, so the value of the neighbor state can be increased. It is assigned the value of the reward multiplied by a discount factor γ. This assignment is called a backup, since we are moving the reward information back up the path.

$$V_s = r + \gamma V_{s'} \quad \textbf{6-2}$$

where the value of this state V_s is the sum of any reward r received, plus the discounted value of the next state $\gamma V_{s'}$.

This form of learning can be unstable since any historical value of V_s is lost. A better approach is to shift the value of the state towards the destination value, so if we go from a high-valued state to a low-valued state we do not lose all memory of that higher value.

$$V_s = V_s + \alpha\left[r + \lambda V_{s'} - V_s\right] \quad \textbf{6-3}$$

where α is the learning rate. Larger values of α provide faster learning, though smaller values of α better preserve the "memory" of the system. Most of the time the reward will be zero, so this serves to diffuse the rare reward information through the states.

Now the non-terminal state s holds a γ-weighted prediction of the reward it can ultimately expect as a result of moving into it.

The model for TD is shown in Figure 6-2. The goal is implied by the rewards returned by the environment.

Since each state has a value, we can take actions preferentially towards the highest valued state at each step. On the next run through the map, the algorithm knows it is beneficial to move to either state + on the map in Figure 6-1. It can tell that these states lead to $*$ and then to the Goal. This greedy search is like traveling uphill, or rolling a ball downhill.

Following a strictly greedy search can stick the system into a rut. It also does not allow for flexibility if the map changes, such as when a door opens as was shown in Figure 5-11. A variation on the greedy search takes a non-optimal action every once in a while. With a probability of ε, we take a random action instead of the best one.

Taking a random action is called exploration, while taking the best action is exploitation. A balance of exploration versus exploitation allows us to improve our map or adapt to change, while still providing an efficient path to the goal state.

A limitation of ε-greedy searching is that we take actions that lead to extremely poor states with the same probability as actions to decent states.

Figure 6-2
TD model

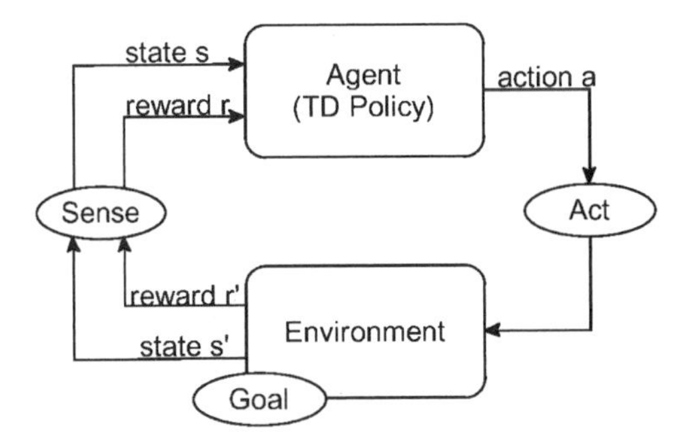

An improvement is to make the probability of taking an action proportional to its expected return. The best action is the most likely one, the second best action is taken less frequently, and really terrible actions are not taken very often. This proportional search is known as softmax. The selection probabilities can be linear or based on an exponential curve, as you prefer.

The entire set of state values is the system policy π. As more and more runs are made across the map, this policy will approach the theoretically optimal policy π^*. Each run through the map is called an episode, and each episode improves the reward predictions, V_s, from the goal outward. It may take hundreds or thousands of episodes to develop a good policy.

This process of following the policy and using the reward information to create predictions of future reward is known as temporal difference (TD) learning. TD learning provides the basis for the rest of the RL techniques. The basic algorithm is:

Initialize the values V_s in policy π
For each episode:
  Choose some start state s
  Until the goal is reached:
    Choose some action a using policy π at state s
    Take action a and observe the reward r and next state s'
    $V_s = V_s + \alpha[r + \lambda V_{s'} - V_s]$
    $s = s'$

While this may seem terribly inefficient compared to the A* search used earlier, the algorithm can be applied to a variety of problems that cannot be solved by A*.

A comprehensive introduction to RL and many of its variants can be found in Sutton (1998).

SARSA

The TD algorithm, using state values V_s, requires that the system be able to look ahead to see the value of neighboring states. In a map situation this may be okay, but when we are trying to control a process it may not be possible, especially if we do not have a good predictive model of the process.

This is fixed by moving the estimated reward value off of the state and into the actions for the state. We then learn the policy for these action-values instead of the state-values. The action-value form of Equation 6-3 is shown in Equation 6-4.

$$Q_{s,a} = Q_{s,a} + \alpha(r + \gamma Q_{s',a'} - Q_{s,a}) \quad \text{6-4}$$

The value Q of action a at state s is adjusted by the reward received from taking this action plus any adjustments made based on the expected reward from taking the policy-defined action at the next state. The value adjustment is made after taking action a and arriving at state s'.

This evaluation rule makes use of all five elements (s, a, r, s', a') involved in the transition from one step to another, hence the name *SARSA*.

The example of a windy grid in Figure 6-3, from Sutton (1998), illustrates both the use of $Q_{s,a}$ instead of V_s and an example where RL can learn a solution where A* may not be able to operate.

On this map, as far as the agent knows, it just has to move to the right to reach the goal. However, in the middle of a map is a "wind" that is not in the agent's model. At each time step as the agent crosses the windy zone, the wind blows the agent up across the grid. Nonetheless, the SARSA algorithm eventually finds the indicated path in spite of this unpredicted interference.

The SARSA algorithm is very similar to the basic TD algorithm:

Initialize the values $Q_{s,a}$ in policy π
For each episode:
  Choose some start state s
  Choose action a from state s using policy π
  Until the goal is reached:
    Take action a and observe the reward r and next state s'
    Choose action a' using policy π at state s'

$$Q_{s,a} = Q_{s,a} + \alpha(r + \gamma Q_{s',a'} - Q_{s,a})$$

$$s = s'$$

$$a = a'$$

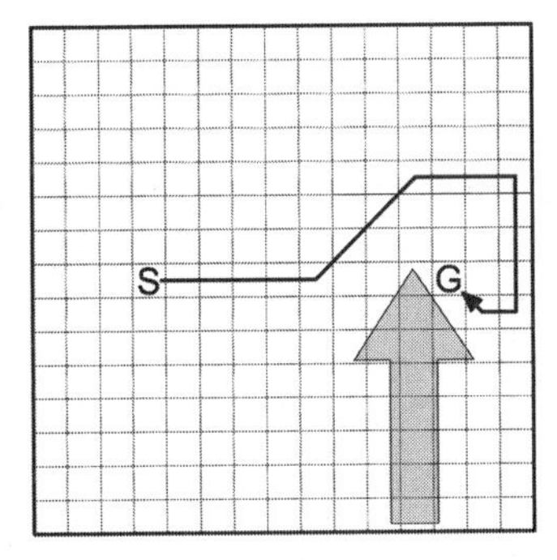

Figure 6-3
Windy grid

Off-Policy Q-Learning

During SARSA learning, the ε-greedy or softmax search does a good job of balancing the needs of exploration and exploitation. However, whenever that search takes a non-optimal action, the value estimate of the previous action can be degraded.

Instead of updating the value of $Q_{s,a}$ based on the action $Q_{s',a'}$, we can update it *as if* we took the greedy action $\max_{a} Q_{s',a'}$. Since we are updating the policy based on an action we did not take, this is known as off-policy learning. When we learn based on the actions actually taken, it is on-policy learning. This variation is shown in Equation 6-5.

$$Q_{s,a} = Q_{s,a} + \alpha\,(r + \gamma \max_{a'} Q_{s',a'} - Q_{s,a}) \qquad \textbf{6-5}$$

Regardless which actual action a' the search takes, our backup value is based on the best action that could be taken according to the current policy.

The Q-Learning algorithm is shown here.

Initialize the values $Q_{s,a}$ in policy π
For each episode:
  Choose some start state s
  Until the goal is reached:
    Choose action a from state s using policy π
    Take action a and observe the reward r and next state s'
    $Q_{s,a} = Q_{s,a} + \alpha\,(r + \gamma \max_{a'} Q_{s',a'} - Q_{s,a})$
    $s = s'$

Eligibility Traces

All of the RL methods introduced so far only provide a one-step return of the reward values. This makes learning a good policy slow, tedious work. However, it is possible to distribute a single reward to the many actions that led up to it. This is done by adding an eligibility attribute $e_{s,a}$ to our actions. This keeps track of the eligibility for each action to receive part of the return from the reward. Whenever an action is performed, its eligibility is updated:

$$e_{s,a} = 1.0 \qquad \textbf{6-6}$$

At each step, all other eligibility values are reduced by a decay factor λ:

$$e'_{s',a'} = \lambda e'_{s',a'} \qquad \textbf{6-7}$$

A reward is then allocated to each state according to their eligibility:

$$Q_{s,a} = Q_{s,a} + \alpha e_{s,a}\left(r + \gamma \max_{a'} Q_{s',a'} - Q_{s,a}\right) \qquad \textbf{6-8}$$

To avoid updating each and every action or state in the system, you can keep a list of states that have an eligibility above some threshold amount. States that decay below this threshold get dropped from the active list.

Equation 6-8 can be made more tractable by adding a new term δ to keep track of the return:

$$\delta = r + \gamma \max_{a'} Q_{s',a'} - Q_{s,a}$$

$$Q_{s,a} = Q_{s,a} + \alpha \delta e_{s,a} \qquad \textbf{6-9}$$

This system maintains a trace of all of the actions that led up to the reward, with actions that occurred farther back in time having a lower eligibility and hence a lower adjustment from the return.

This speeds up the process of learning the optimal policy.

Though we demonstrated eligibility traces for Q-Learning, they can be applied to all forms of RL.

The algorithm for eligibility traces in Q-Learning is shown here.

Initialize the values $Q_{s,a}$ in policy π
Set all eligibility $e_{s,a} = 0$
For each episode:
    Choose some start state s
    Until the goal is reached:
        Choose action a from state s using policy π
        Take action a and observe the reward r and next state s'
        $\delta = r + \gamma \max_{a'} Q_{s',a'} - Q_{s,a}$
        $e_{s,a} = 1.0$
        For all (s^*, a^*) with non-zero eligibility:
        $Q_{s,a} = Q_{s,a} + \alpha \delta e_{s,a}$
        $e'_{s',a'} = \lambda e'_{s',a'}$
            $s = s'$

Code: Continuous Control

To apply RL to a dynamic control problem, it is necessary to map the continuous state parameters into a discrete form usable by the RL algorithm. Let us look at this problem in regards to the classic inverted pole on a cart problem shown in Figure 6-4.

The pole–cart is one of the simpler physical simulations. Others, including one that learns to stand up by Jun Morimoto and Kenji Doya, can be found in the literature and on the Internet.

The state of the cart is described by the angle θ of the pole, the position x of the cart, and the rate of change, or velocity, of both these values.

There are only three actions available to the agent; push the cart left, push the cart right, and let the cart coast.

The cart's motion is described in Equations 6-10 through 6-14.

$$\beta = \frac{F + m_p l \dot{\theta}^2 \sin\theta}{m_c + m_p} \tag{6-10}$$

$$\ddot{\theta} = \frac{g \sin\theta - \beta \cos\theta}{l\left(\frac{4}{3} - \frac{m_p \cos^2\theta}{m_c + m_p}\right)} \tag{6-11}$$

$$\ddot{x} = \beta - \frac{m_p\, l\ddot{\theta} \cos\theta}{m_c + m_p} \tag{6-12}$$

$$\dot{\theta} = \dot{\theta} + \tau\ddot{\theta}$$
$$\dot{x} = \dot{x} + \tau\ddot{x} \tag{6-13}$$

$$\theta = \theta + \tau\theta$$
$$x = x + \tau\dot{x} \tag{6-14}$$

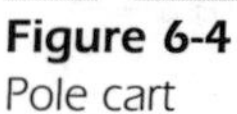
Figure 6-4
Pole cart

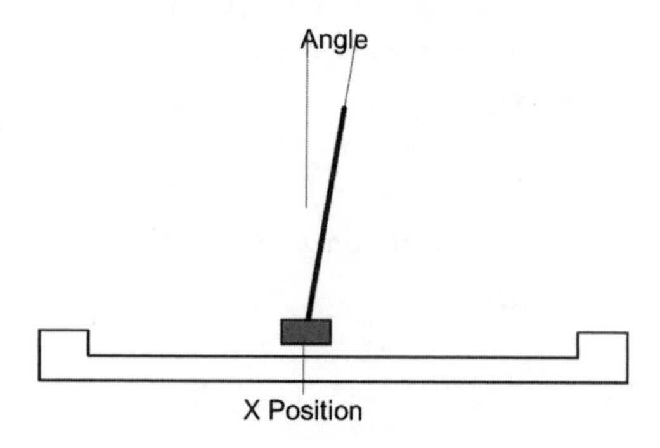

where:

F is the force exerted on the cart
x is the position of the cart
θ is the angle of the pole on the cart
l is the half the length of the pole in meters
m_c is the mass of the cart in kilograms
m_p is the mass of the pole

$\dot{x}$, is the velocity of the cart
$\dot{\theta}$ is the velocity of the pole's angle
$\ddot{x}$ is the acceleration of the cart
$\ddot{\theta}$ is the acceleration of the pole's angle
τ is the time step, required by Euler's method of simulation
g is the acceleration from gravity, 9.8 m/s^2.

Reaching into the RL bag of tricks, let us try to learn to balance the pole using Q-Learning with eligibility traces. Since this problem is continuous and not tabular (discrete), there are some unique problems to address.

The system state is not an index into a map but a set of continuous variables. These four continuous parameters can be partitioned, or quantized, into indexes into a four-dimensional array. For large dimensions or detailed quantizations this state array can become very large.

The implementation of this comes together in the program *RLCart*. Figure 6-5 shows the class diagram for *RLCart* and its supporting packages.

This system is implemented across two packages. *RL*, *Agent*, *Environment*, and *Sense* make up the base RL packing in aip.rl. The application specific objects in aip.app.rlcart are *RLCart*, *PoleCart*, and the specific learning agent *AgentBoxesQ*.

Figure 6-5 RLCart class diagram

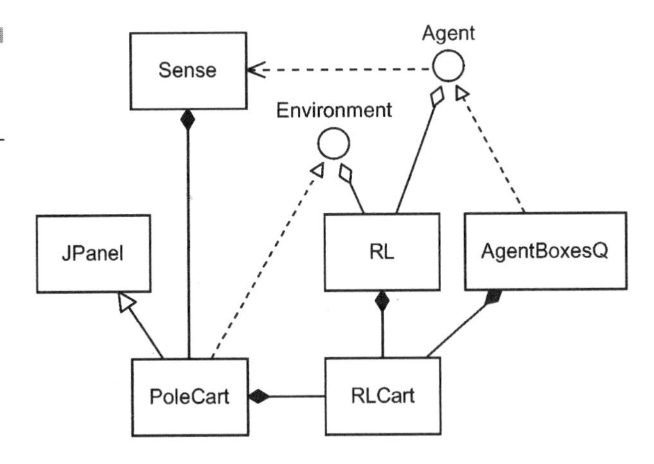

Table 6-1

Sense summary

```
Construction
            Sense (int dim)

State
       int  getDim()
  double[]  getVector()
      void  setVector (double[] vector)
      void  setValue (int idx, double val)
    double  getValue (int idx)

Reward
    double  getReward ()
      void  setReward (double reward)
```

Sense.java

aip.rl.Sense The Sense object provides a standard form of communication between the Environment and the Agent in the RL system. It is an array of floating point values plus a reward.

A summary of *Sense* is given in Table 6-1.

Sense(int dim) The dimension of the *Sense* vector is fixed at creation.

getVector()
setVector(double[] vector)
getValue(int idx)
setValue(int idx, double val) These methods provide access to the state represented by *Sense*.

getReward()
setReward(double reward) These provide access to the reward that the environment incurred.

Environment.java

aip.rl.Environment *Environment* defines a standard interface for application specific simulations. There are two methods defined for environments.

start() initializes the environment and returns the *Sense* vector for the start state.

step(int act) performs the indicated action in the simulation and returns the subsequent *Sense*.

Agent.java

aip.rl.Agent The *Agent* is much like the *Environment*, providing a simple framework on which to hang a learning algorithm.

Its *start(Sense state)* method takes the starting state of from *Environment* and uses it to initialize itself.

step(Environment environ) takes a step in the environment and uses the resulting information to learn how to step more efficiently in that environment.

AgentBoxesQ.java

aip.app.rlcart.AgentBoxesQ This specific agent implements the *Agent* interface to provide actor/critic learning with a specialized state space quantization for this application.

PoleCart.java

aip.app.rlcart.PoleCart The *PoleCart* class implements the *Environment* as a pole and cart simulation. It is an extension of *JPanel*, so it has a place to draw its cart and pole. It also takes a *JLineGraph*, so it can display its internal state as a phase diagram (x vs. $\dot{x}$, and θ vs. $\dot{\theta}$).

The result of its *start()* and *step()* methods is the *Sense* of the environment at that point in time.

RLCart.java

aip.app.rlcart.RLCart *RLCart* is a *JFrame*-based application that provides the UI for *PoleCart* and a history of each learning episode.

RL.java

aip.rl.RL The *RL* class provides a control wrapper for the other RL classes.

A summary of *RL* is given in Table 6-2.

Table 6-2
RL summary

Construction	
	`RL (Environment environ, Agent agent)`
`void`	`start ()`
Operation	
`void`	`runSteps (int steps)`
`void`	`runEpisodes (int episodes, int steps)`
Access	
`Environment`	`getEnvironment()`
`Agent`	`getAgent ()`
`int`	`getStep ()`

Representing State Space

This section looks at different ways to represent a map, and by extension any arbitrary state space, and the values that guide navigation through it.

Map Decompositions

In the section on A* *Path Finding*, state space was a convenient grid of positions on a map. Each position was either legal or illegal to step into. Finding a path through this grid was simply a matter of calculating the shortest distance from the start to the end.

However, a grid can be an inefficient way to represent a large area. Other representations include quadtree decomposition, binary space partitions, and exact polygonal representation. Figure 6-6 shows these different decompositions of the map.

Quadtree partitioning performs a recursive grouping on the grid. Where there is a square of four matching tiles they are combined into one larger tile. This merging is repeated until it can go no further. Quadtrees are used extensively in digital image processing and computer graphics, as well as for computer games. Additional information can be found scattered all through the literature.

Binary space partitioning (BSP) is another recursive method. At each step the map is split into two parts along a natural boundary in the map. The resulting data structure resembles a binary tree. BSP was brought to the foreground in games through its use in Doom and is a staple for interactive graphical environments. BSP is also used in constructive solid geometry (CSG) algorithms and during Boolean operations on polygons.

Figure 6-6 Map decomposition

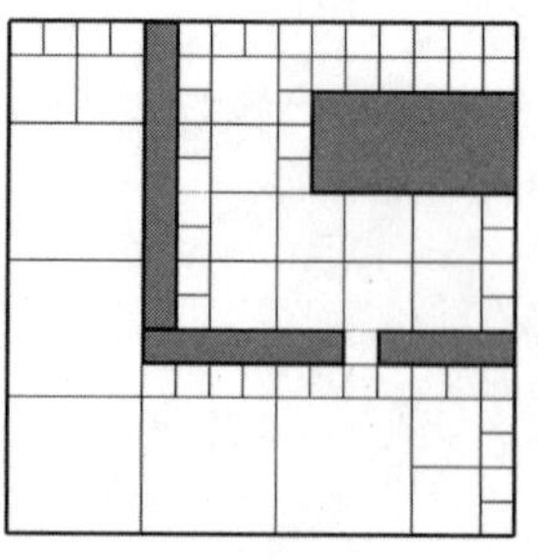

Quadtree

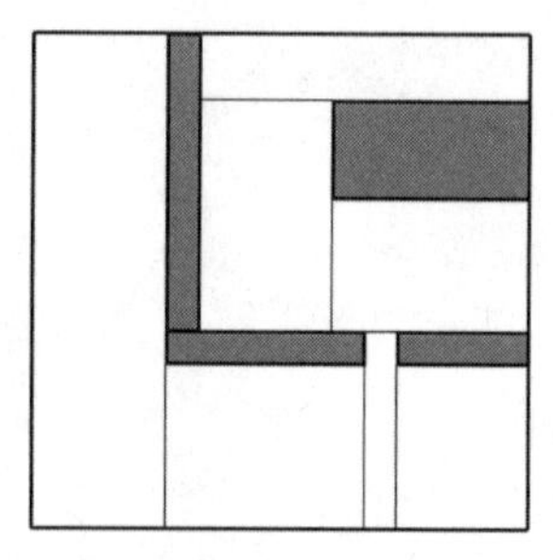

Binary Space Partitioning

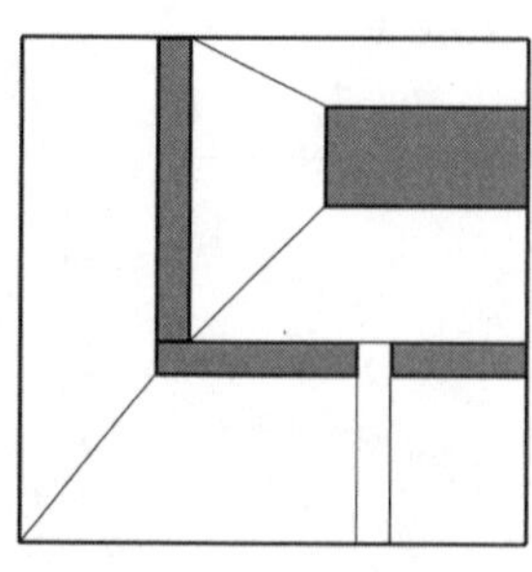

Exact

The exact or arbitrary polygon (or, in this case, quadrilateral) method divides the space up by splitting it at natural features such as corners. Depending on the needs of your application this can be done with arbitrary polygons, quadrilaterals, or triangles. You can find references for decomposition into triangles (and sometimes other forms) in references on computational geometry as well as computer graphics.

Partioning Continuous Space

The section on *Reinforcement Learning* started with a discrete map, where each grid entry has a value. An agent navigating that map would move to the highest valued neighboring node. A variation associated the value with the actions leading from the node instead of the node itself, with the agent taking the highest rated action.

When the map went from a grid to a continuous space, we were given the option of simply quantizing it back into a grid.

Tile Space Most systems will not visit every possible state, so it is possible to take advantage of this sparse coverage to reduce the size of the state storage. The quantized state vector can be hashed to fit into a smaller array.

A pure quantization, where each state maps to one partition, or tile, can learn the function but it does not leverage its experience in one state to kick-start a neighboring state. Two neighboring states will tend to have similar reward expectations, and this continuity could be used to advantage. With discontinuous tiles, the experience of one state does not generalize to any neighboring states at all.

Instead of placing a state into a tile on just one grid, it could be placed on tiles in several grids. Each of these grids would be offset from each other as shown in Figure 6-7. On some layers, two states will share a tile and learn from each other's rewards while on other layers they may be in neighboring tiles.

These overlapping layers of tiles allow the system to learn a continuous value function that generalizes across neighboring states. The set of tiles that overlap a given state are known collectively as the feature for that state.

I discovered this concept in Sutton (1998) who, it would seem, adapted it from a system that simulates a receptive field for neural network robotic control at the University of New Hampshire (`www.ece.unh.edu/robots/cmac.htm`).

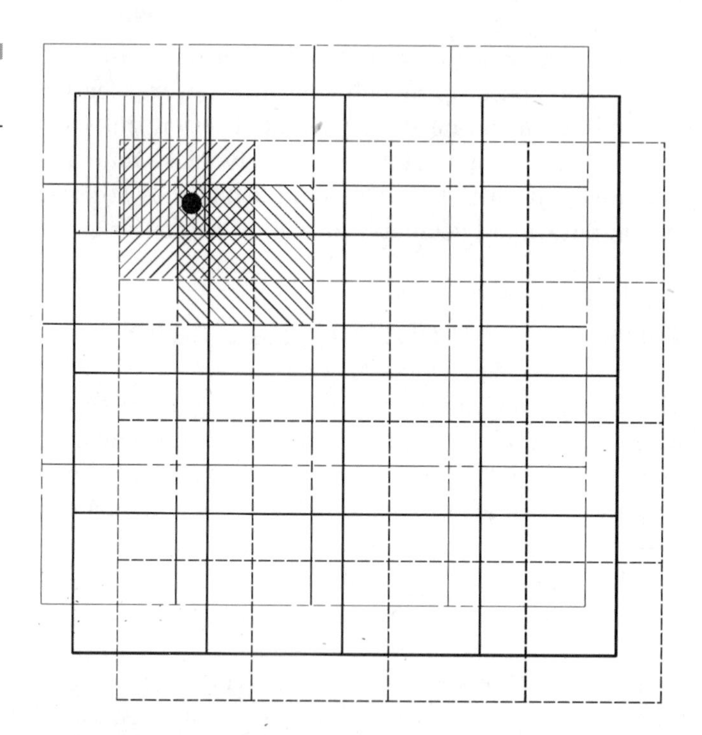

Figure 6-7
Tile space

This tile space coding solves the state quantization problem, provides for generalization, and also gives us the basis to solve our other issue: how to learn the rewards for different actions in state space.

In the tabular form it was possible to simply look in a neighboring box to see what a reward was. With continuous linear state parameters we also have continuous linear expected rewards. It becomes necessary to learn the continuous reward space at the same time as we learn how to navigate in it.

The overlapping layers of tiles are used to learn the reward space using gradient descent techniques borrowed from neural networks.

However, the pole–cart system is simple enough to use simple parameter quantization, without tiling.

This implementation uses the format known as actor/critic. These type of systems split the learning into two parts.

The actor learns which actions are best for a given position in state space, and the critic learns the predicted value of a position in state space. The actual value is checked against the predicted value, and this error is used to train both the actor and the critic. The critic's estimate of a state's value is used to weight the learning done in that state; the critic kibitzes the actor.

The actor is the RL system as we have learned about it so far. The critic learns the state space parallel to the actor. The critic affects the learning weight that is used by the actor. The algorithm is still essentially like Q-Learning, with parallel learning systems.

Initialize the weights and traces to zero
For each episode:
    Choose some start state s
    Until the goal is reached or failure:
        Choose action a from state s using the actor policy
        Set the actor trace e_a to the action taken
        Set the critic trace e_c
        Take action a and observe the reward r and next state s'
        Get the reward weight w_t from the critic's $W(s)$

$$\delta = r + \gamma w_t - w_{t-1}$$
$$Q_s = Q_s + a\delta e_{\alpha,s}$$
$$W_s = W_s + \alpha\beta\delta e_{c,s}$$
$$e_{a,s} = \lambda_a e_{a,s}$$
$$e_{c,s} = \lambda_c e_{c,s}$$
$$s' = s$$

Occupancy Grids The A* grid of obstacles is a kind of occupancy grid, where obstructed areas are blocked off. In the real world of robots, sensory data rarely gives us such a clean picture of the environment. Researches such as Johann Borenstein, currently at the University of Michigan, blur the occupancy grid to account for the uncertainty inherent in sensor data. This gives a histogram map, where evidence is accumulated through repetition.

An area of the map, on first scan, might be lightly "shaded" with occupancy evidence but later, if the robot gets a better scan of it or even bumps into something there, the evidence can be upgraded. This provides a shades-of-grey grid. The more obstructed an area is on the map, the higher its cost may be. An agent navigating that space will tend to stay in clear, unobstructed areas.

An alternative visualization puts the obstruction value as a height above the grid. The agent then navigates along the valleys of the terrain.

Vector Fields The statistical occupancy grid stores its information in the grid locations and acts like a height field. Vector fields store their information as actions, or preferred actions, across the field and act like electric charges or repulsive fields that guide the agent.

A potential field representation applies a "charge" to each obstacle in the environment. The agent navigating the environment is repulsed by these charges and the closer it gets to an obstacle the stronger the push. The repulsion can be represented as acceleration vectors pointing away from the obstacles.

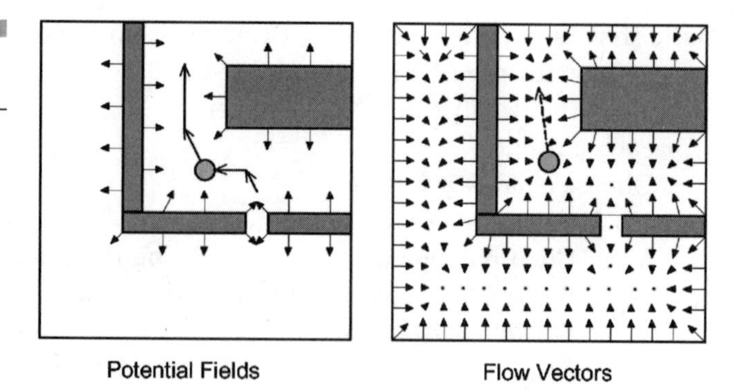

Figure 6-8
Vector fields

The path of least resistance is found by summing all of the obstacle accelerations with the vector that represents the direction the agent wants to go. The larger the agent's direction vector the more it will override the bumps and wrinkles imposed by the environment.

In closed areas like doorways the opposing repulsion vectors cancel themselves out near the center, providing a clear path.

Instead of storing repulsion information with the obstacles and then summing across all of the obstacles, this information can be kept as preferred action vectors across the environment. The effect is the same, but most of the work can be done with pre-calculations, or with occasional updates to the map. Navigation is then done using local information. These ideas are illustrated in Figure 6-8.

The flow vectors do not have to be calculated based on object repulsion. The vectors can represent any force needed to guide the agent, including object repulsion, goal attraction, etc.

Another way to think about vector fields is as a flow of water instead of a repulsion from obstacles. The agent is simply riding the vector "current" from one place to another. While the vector field would be represented the same way, the vectors could be calculated using various techniques.

Genetic Algorithms

The various forms of genetic algorithm (GA) find an optimal or near-optimal solution to a problem by, essentially, wandering around at random in state space. While this sounds horribly inefficient, there are a few refinements that make it work.

The search is done in parallel, even if that is only an illusion, by a large population of agents. These agents in the search space are evaluated according to how well they are solving the problem so far. The best performing agents are kept while the worst ones are replaced by new ones created by combining traits from existing, well-performing agents.

This way, information about good areas in state space is retained in the form of surviving agents. That information, in the form of a genetic code, is shared between agents when they combine to create new agents.

Genetic thinking was first applied to computer problems by John Holland, who adapted and refined what we now know as Genetic Algorithms during the 1960s and 1970s (Holland, 1975).

Computational GA follows the model developed by nature. Each gene encodes some trait, and these genes are collected into chromosomes. The set of chromosomes that define an individual entity are its genome, and a specific instance of a genome is that entity's genotype. The form the entity takes from interpreting the genotype, such as a specific solution to a problem, is its phenotype.

A large population of these entities is created and each entity is evaluated as to how well it solves the problem. This defines their fitness, or how worthy they are of becoming the final solution to the problem.

Unfit entities are removed from the population and the fit entities breed to replace them. During the breeding process, information is shared between entity pairs with crossover and other genetics-based operations. Errors can also be introduced through mutation. The resulting offspring are then evaluated along with their parents and the cycle continues.

Once the genetic diversity of the population converges to a group of highly related individuals, the problem is considered solved and the most fit entity is used as the solution. Optimality through inbreeding.

```
Generate a random population of entities, or agents
Evaluate the fitness of each agent
Breed a new population:
    Select pairs of agents according to their fitness
    Combine the parents to create additional agents with traits of
    both parents
    Apply a small amount of random mutation
Repeat until done
```

Encoding the Problem

While the GA itself is very simple, finding a way to map the problem you want to solve into mutable chromosomes can be tricky. In fact, finding a good encoding for the problem is one of the most important steps in using GA.

For a problem to be solvable by GA methods, it must first be converted to parametric form. There must be some set of parameters that can define the solution to the problem. There must also be some fitness function that determines the quality of the solution those parameters define.

A variation of GA is genetic programming (GP), which evolves program code that is then executed. The result of the execution determines the fitness of the agent.

Different forms of encoding and how they are affected by mutation are explored here. The process of breeding and mutation is discussed later.

Value In value-based coding, each gene is treated as a single unit from some alphabet of possible units. Examples of values include characters, numbers, and enumerations. Each gene could be anything from a character to a complex data structure.

An example enumeration in the domain of color might be:

Chromosome: `[red] [blue] [red] [black] [green]`

where the possible colors are [black], [red], [orange], [yellow], [green], [blue], [purple] and [white].

A mutation would bump the color up or down the list to a neighboring value or it could randomly replace the color with one from the alphabet.

Another enumerated form uses a subset of the ASCII characters in a string.

A chromosome of real-valued numbers is similar:

Chromosome: `1.23; 4.97; 2.38; 0.92; -4.01`

Mutation involves adding a signed, random number to the existing value. The mutation may be in a fixed range of [–1.0 .. 1.0], scaled according to the target number, or be a predefined offset.

A more flexible approach to real-valued coding is explored later.

Binary The original and, perhaps, most common coding is binary representation of the parameters. Typically, the genes will represent integral values, though it is possible, if not desirable, to apply binary coding to floating point numbers.

First, a brief introduction to the binary number system.

The numbers we use in daily life consist of 10 digits. The digit "1" represents a single thing, such as a pebble. "2" is a pebble next to another. Add another pebble and you can represent their quantity with the symbol "3". The digits are just symbols that stand-in for a particular quantity of pebbles. "0" is special since it represents no pebbles.

Nine pebbles are the most we can represent with a single digit. What happens when we add another pebble? We get the symbol sequence "10".

Each digit's value is affected by its position in the sequence. The rightmost digit is in the "ones" place, and is worth its face value. The next digit to the left is in the "tens" place, where a "1" stands for 10 pebbles, "2" is 20, and so on.

This system is known as the decimal system, or base-10. Each place value is 10 times, the place value to its right, since we have 10 digits. To make the obvious complicated, the quantity represented by the sequence of digits is calculated by:

$$q_{10} = \sum_{p=0}^{n-1} d_p 10^p \tag{6-15}$$

where

q_{10} is the quantity represented by the sequence of base-10 digits
n is the number of digits in the sequence
p is the place number of a digit, from zero at the right to n–1 at the left
d is the quantity represented by the digit at position p

Binary coding is exactly the same but with only two digits, "0" and "1":

$$q_2 = \sum_{p=0}^{n-1} d_p 2^p \tag{6-16}$$

So the binary sequence:

1101011

Represents the quantity:

$$\begin{aligned}(1*2^6) + (1*2^5) + (0*2^4) + (1*2^3) + (0*2^2) + (1*2^1) + (1*2^0)\\ = 64 + 32 + 8 + 2 + 1\\ = 107\end{aligned}$$

All numbers inside your computer are represented by sequences of binary digits, also known as bits. Bytes are eight bits and can represent the decimal values of +127 to −128 or the values 0 to 255, depending on interpretation. Integers are normally represented by 16 or 32 bits.

A binary mutation flips a single bit from 0 to 1 or from 1 to 0. For example:

Chromosome: `1101011 = 107`
Point Mutation: `1101111 = 111`

Gray Code While binary code is an easy, and traditional, way to represent integers in GA, it also has some problems.

GAs work better when the encoding follows the principle of causality, where a small change in the genotype creates a small change in the phenotype.

In a binary chromosome, a one-bit change in a large-valued position can have a huge change, such as this mutation:

Chromosome: `1101011` = 107

Point Mutation: `0101011` = 43

Another problem with binary coding is that, in some cases, the step to the next highest value requires flipping most of the bits in the code:

Chromosome: `0001111` = 15

Next value: `0010000` = 16

One measure of distance in a binary number is the Hamming distance. This is defined as the number of bits that are different between two binary sequences. The step from *107* to *43* has a Hamming distance of one, but a value difference of *64*, for poor causality. The Hamming distance between *15* and *16* is five but with a value difference of one. This represents the situation known as a Hamming cliff.

Gray coding provides a solution to the Hamming cliff at least. Bell Labs researcher Frank Gray gave his name to the system with his patent #2,632,058 for "Pulse Code Communication" in 1953.

In Gray code, each value is separated from its neighboring value by a single bit flip. This fixes the Hamming cliff problem. For example, the first 16 Gray codes and their value and binary equivalents are given in Table 6-3.

Converting from binary to Gray code is simple:

$$G = B \wedge (B/2) \qquad \textbf{6-17}$$

where "^" is the exclusive-OR operation:

$$\begin{aligned} 0 \wedge 0 &= 0 \\ 0 \wedge 1 &= 1 \\ 1 \wedge 0 &= 1 \\ 1 \wedge 1 &= 0 \end{aligned}$$

Unfortunately the conversion back to binary is not as simple, and must be described as an algorithm.

$B = G$ **6-18**

For an N-bit number, repeat N times:

$B = B \wedge (B/2)$

Table 6-3

Gray codes

	Binary	Gray
0	0000	0000
1	0001	0001
2	0010	0011
3	0011	0010
4	0100	0110
5	0101	0111
6	0110	0101
7	0111	0100
8	1000	1100
9	1001	1101
10	1010	1111
11	1011	1110
12	1100	1010
13	1101	1011
14	1110	1001
15	1111	1000

Repeating our mutation example in Gray code, we get:

Chromosome: `1011110` = 107

Point mutation: `0011110` = 23

The left bits in a Gray code still have exponentially more value than the right bits, so a single bit flip can still have a large affect on the parameter value.

We can still strive for the next level of causality, where a small change in the chromosome's value causes a small change in the phenotype. The GA will simply have to learn not to mess with the high-valued bits.

The Hamming cliff is gone, at least. Looking at Table 6-3, there are no adjacent codes with more than one bit of difference. From our previous example:

Chromosome: 15 = `0001000`

Next value: 16 = `0011000`

Real-valued The binary representation of the floating point numbers used by your computer can be used as GA chromosomes. This is not necessarily the most efficient way to handle floating-point numbers in GA.

Another technique is to use value encoding on the real number. Mutations are then fixed-step adjustments to the value. A mutation with a fixed step size of δ would be:

$$V = V \pm \delta$$

or the step could be a scaled random value:

$$V = V \pm \delta r$$

where r is a random number between 0 and 1.

For that matter, the step could be based on the value's current size:

$$V = V \pm \delta V$$

A more common method of handling real-valued chromosomes is to represent them as integers in the gene and then scale them to match the parameter's needs during the decoding:

$$V = a + (b - a)\frac{I}{2^n - 1} \qquad \textbf{6-19}$$

where

V is the real value decoded from the integer
a is the lowest acceptable value
b is the highest acceptable value
I is the integer value from the gene
n is the number of bits used by the gene.

This system only works when you have a good idea of the range for a and b. Dynamic systems analyze how the gene is used across the population and can perform self-scaling; so the GA focuses on the useful range of the parameter and does not spend time optimizing in irrelevant areas of state space. A discussion of dynamic parameter encoding can be found in Schraudolph (1992).

Permutation Some problems have special requirements. The traveling salesman problem (TSP), for example.

In the TSP you are given a list of cities on a map. The goal is to find the shortest path that visits all of the cities exactly once.

The chromosome consists of a list of all the cities in a particular order. The fitness of the chromosome is the length of the path that visits the cities in that order.

Each chromosome must contain each city exactly once, so random mutation of cities is clearly out of the question. The various crossover operations must also guarantee the correctness of the chromosome.

Tree While most GA encodings operate on an ordered set of genes, it is also possible to evolve trees. Mutations could add or remove nodes or sub-branches from the tree, and crossover would split the two parent trees at some arbitrary positions and swap the sub-trees. These operations are illustrated further in the next section.

What is interesting about trees is that they can represent computer programs. Each node is an operator or function and each leaf is a constant or variable. For example, Equation 6-17 expressed as a tree is shown in Figure 6-9. With the addition of programming constructs such as *if* and *for*, complex programs as well as formulas can be evolved.

GAs applied directly to program code rather than the parameters that control an existing function are called genetic programming (GP). GP was introduced by John Koza in 1989, and he now has a series of books on the subject.

The genetic alphabet in GP consists of all relevant functions plus all relevant leaf nodes. Its fitness is determined by running the program and observing how well it performs in the problem space.

The alphabet used for GP must respect the properties of closure and sufficiency.

Closure means that all of the functions can accept as their operators all of the other entries in the alphabet, including the return value of other functions.

Sufficiency means that the chosen alphabet is capable of expressing the solution to the problem.

Interestingly enough, a somewhat involved system of notation can bring the representation of the tree back into a string coding. L-systems, named after Aristid Lindenmayer, describe hierarchical branching structures. L-systems were developed to emulate plant and cell growth in

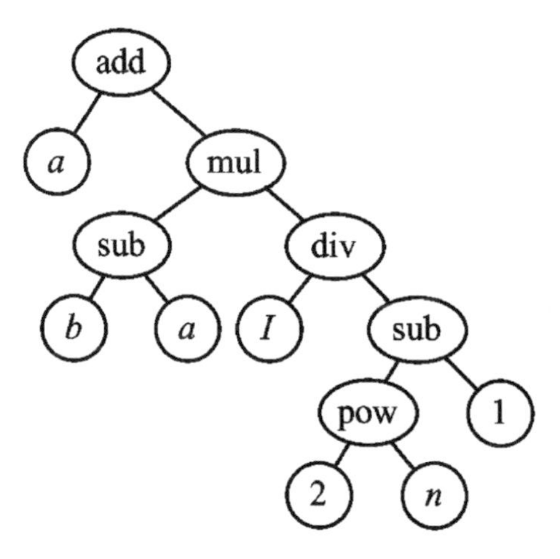

Figure 6-9
a + (b-a) (1/(2^n-1))

nature, but their concepts could apply to GP as well. Conversely, tree-based GP could be used to evolve the normally string-based L-systems.

An excellent introduction to L-systems is found in Prusinkiewicz (1991).

The alphabet for L-systems include a push "[" and pop "]" operator. These are used to define "branches" in the tree. For example, a template string for addition would be:

```
+[ ][ ]
```

where the first branch is the left-hand side of the addition, and the second branch the right-hand side.

Of course, the mutation and crossover rules would have to respect the branch boundaries or the decoder would need to be able to handle unbalanced push and pop operators.

Equation 6-17 expressed in this notation would be the frightening string:

```
+[a][*[-[b][a]][/[I][-[^[2][n]][1]]]]
```

Breeding

Once the problem has been encoded into some form of genome, the algorithm aspect of GA can take over. There are two steps in the breeding process, crossover and mutation.

Crossover is the primary force behind breeding. This is where the genotypes from both parents are combined to create one or more child genotypes.

During the breeding process there is a small chance of an error in the reproduction of the genotype. This introduces a change in the chromosome, a spontaneous mutation. Mutations provide the kick needed to help the GA explore state space. However, too high of a mutation rate turns the orderly evolution of the agents into a random scramble.

Crossover To create offspring, the genotypes of two parent entities must be combined. Taking chunks from one parent and combining them with chunks from the other parent create a new, different entity. This process is called crossover because it looks like the parent genomes cross each other, as shown in Figure 6-10.

If the two parent chromosomes are defined as these strings:

Parent 1: `ABCDEFGHI`
Parent 2: `123456789`

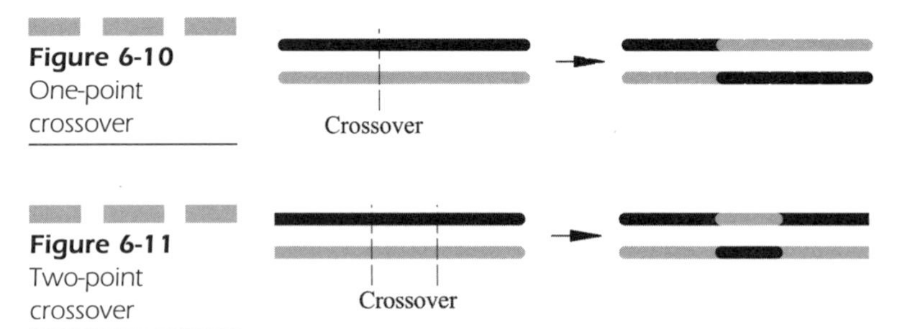

Figure 6-10 One-point crossover

Figure 6-11 Two-point crossover

With a crossover between the fourth and fifth entries, we get two possible children:

Child 1: `ABCD`**`56789`**

Child 2: `1234`**`EFGHI`**

Two-point crossover chooses two random positions to split the genotypes (Figure 6-11). There can be any number of crossover points, up to uniform crossover where each chromosome is selected at random from one parent or the other.

Permutation In permutation codings, simple crossover operations violate the rule that all entries appear exactly once in each agent. For example, given the two-point crossover at the "!" marks:

Parent 1: `BGA!FCE!HD`

Parent 2: `BGF!EAC!HD`

Child 1: `BGA`**`EAC`**`HD`

Child 2: `BGF`**`FCE`**`HD`

The two children are no longer proper permutations of the alphabet.

Instead of swapping sections, one form of permutation uses the crossover selection as a mapping between the two chromosomes. The example above specifies the crossover mapping:

`F` to `E`

`C` to `A`

`E` to `C`

The children chromosomes then begin life as exact copies of their parents:

Child 1: BGAFCEHD

Child 2: `BGFEACHD`

The chromosomes then are scanned and the genes are redefined according to the map. Note that the mapping goes both ways. The original `F` is mapped to an `E`, and the original `E` is mapped to an `F`, swapping the genes:

Child 1: `BGA`**`E`**`C`**`F`**`HD then BG`**`CEA`**`FHD then BG`**`EC`**`AFHD`

Child 2: `BG`**`EF`**`ACHD then BGEF`**`CA`**`HD then BG`**`C`**`F`**`E`**`AHD`

Note that you get slightly different results if you perform the swaps simultaneously or one at a time.

Tree Single-point crossover in tree genomes operates on the same principle as in string genomes. A point is selected in both parents where the genome is split. The children are composed of one half of the first parent and the other half of the other parent. This is illustrated in Figure 6-12.
The programs:

```
a+((b-a)*(I/2))
(I/2)/n^2
```

Cross over to create the programs:

```
a+((b-a)*n^2)
((I/2)/(I/2))
```

Adaptive Chromosomes We challenge two assumptions about the chromosome here. The first assumption is that there is a fixed number of genes in a chromosome and a fixed number of chromosomes in the genome. Markers provide a way of evolving these parameters, and can be used in applications such as growing neural networks.

The other assumption is that the crossover operation can occur just anywhere on the genome. Adaptive crossover allows us to evolve optimal crossover spots in the genome.

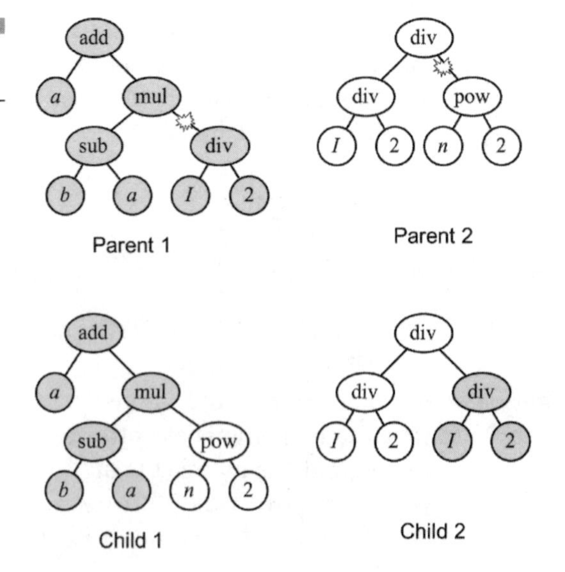

Figure 6-12
Tree crossover

Markers are special entries in a chromosome that indicate the start and end of a sequence of genes. Each sequence between start and stop may define some module in the problem space, such as a layer in a neural network. The push and pop markers for L-systems are a type of start and stop marker.

The ability to evolve flexibly sized chromosomes can help in certain hard problems, where it is not possible to determine the best coding for the problem in advance.

The use of chromosome markers is explored in, for example, Moriarty (1993) and Fullmer (1991).

The process of evolving the crossover points instead of choosing them randomly was pioneered by Schaffer (1987). Evolving the crossover positions allows the evolutionary process to couple genes together, so they travel together. This can be beneficial if the fitness of the organism depends on the relationship between a set of genes.

Adaptive crossover requires a parallel string of bits that evolves along with the data-bearing genome. For example:

```
Parent 1:     ABCDEFGHI
Templatel:    1000010000
Parent 2:     123456789
Template 2:   010001010
```

These crossover points are interpreted as punctuation on the parent genome. The punctuation is attached to a gene and is considered a part of it. The parents, with the crossover punctuation, look like:

```
Parent 1:  AB CDE!F GH I
Parent 2:  12!345 6!78!9
```

Since the punctuation is attached to a gene, they crossover with the genes. Mutations apply to the crossover bits. The two children that are the result of this breeding example are:

```
Child 1:  AB345F78!I
Child 2:  12!CDE!6!GH9
```

Note that, depending on the distribution of the crossover marks, one child may get most or all of the marks. Of course, a child that does not perform well will be eliminated from the population and its poorly developed punctuation will leave with it.

Mutation

Where crossover provides a way to blend the genetics of two parents, mutation adds a touch of randomness. While the offspring will still be mostly like their parents, they will also be subtly different.

All the mutations take the crossed-over chromosome and change one or more entries in it. Different mutation operations are available, depending on the genes in the chromosome.

A point mutation in a value-based chromosome may replace an entry with a random entry from the alphabet. If the values are numbers, one number may be incremented or decremented randomly:

Chromosome: `[red] [blue] [red] [black] [green]`
Mutation: `[red]` **`[green]`** `[red] [black] [green]`
Chromosome: `1.23; 4.97; 2.38; 0.92; -4.01`
Mutation: `1.23; 4.97;` **`2.37`**`; 0.92; -4.01`

For the various binary-code chromosomes, a point mutation can simply flip a single bit using the XOR operator:

Chromosome: `1101011`
Mutation: `1101`**`1`**`11`

In a 32-bit genome, this line of code will flip a random bit, from bit 0 through bit 31:

```
C = C ^ Math.pow(2,(int)(Math.random()*31));
```
6-20

In chromosomes that allow a variable number of genes, such as trees, marker-based chromosomes, and others, you can apply mutations that add or remove genes.

An insertion mutation will add a random gene at some point in the chromosome:

Chromosome: `110!1011`
Insertion: `110`**`0`**`1011`

A delete mutation removes a bit or value from the chromosome:

Chromosome: `[red] [blue] [red]` **`[black]`** `[green]`
Deletion: `[red] [blue] [red] ! [green]`

Random adjustments do not work well for permutation-based chromosomes, so mutations in those tend to re-order genes instead of changing them directly (Buckland, 2002). These shuffling mutations can also be applied to non-permutation chromosomes to varying effect.

An exchange mutation selects two genes in the chromosome and swaps them:

Chromosome: `AB`**`C`**`DE`**`F`**`GHI`
Exchange: `AB`**`F`**`DE`**`C`**`GHI`

A displacement mutation simply shifts genes within the chromosome. It may shift one gene, or select a range of genes to move:

Chromosome: `AB!CDE`**`FG`**`HI`
Displace: `AB`**`FG`**`CDEHI`

Finally, an inversion mutation reverses the order of two or more genes:

Chromosome: `ABCDEFGHI`

Inversion: `ABCDHGFEI`

The inversion, displacement, and exchange mutations can radically change the meaning of the chromosome. They have another application, however, other than changing the meaning of the chromosome. They can be used to pair genes so that they are more likely to survive crossover together, preserving their relationship.

To do this, however, the order of the genes has to be made irrelevant. To do this, the order information needs to be permanently associated with the gene:

Chromosome: `(1,1.23)!(2,4.97)(3,2.38)(4,0.92)(5,-4.01)`

Displace: `(1,1.23)(4,0.92)(5,-4.01) (2,4.97)(3,2.38)`

The values or, more importantly, the meanings of the values, have not changed. Genes one and four are now paired and are more likely to survive crossovers together, the way genes four and five moved together in the example.

Mutations for trees are essentially the same as the mutations for the other types of chromosomes.

Evalutating Fitness

Like all of the other search methods examined so far, GAs require some way of evaluating how close a given solution is to the ultimate goal. In A* path finding this was the distance traveled plus the distance to the goal. The alpha–beta game used a board score, and RL had to interpolate a given state's fitness based on what it ultimately led to.

In GAs, the quality of a solution is known as its fitness. The raw fitness score associated with a genome depends on how well it fulfills its purpose. For a simple mathematical equation, fitness can be calculated directly. For other systems, such as evolving neural networks or strategies for some interactive environment, the agent may need to be run several times and have some sort of score accumulated for it.

Simulated Annealing The genome describes the agent's phenotype, or appearance and behavior, which in turn has a particular fitness in its target environment. As the genome evolves, the fitness of the agent also changes. These fitness changes describe a fitness landscape. For two genes, you get a two-dimensional landscape where the height dimension is the fitness of the agent at those parameter values.

If the fitness landscape is particularly lumpy, the GA can get fixated on a local hill and ignore the towering mountain elsewhere in the map. The goal, however, is to find the highest peak on the map, not just any peak. There are different ways to avoid these local optimums and encourage the GA to find the global optimum.

One method is through simulated annealing. The theory behind simulated annealing is similar to that of regular annealing. The process is run at a high "temperature" at the start, where the randomness, or mutation, component is fairly high. This promotes a lot of broad exploration of the fitness landscape. However, with a high temperature the system never settles down to a solution; it keeps getting bumped around.

So the temperature, or randomness, is slowly turned down until it is gone. The search slowly stops jumping around and settles down on the various peaks in the landscape. This raises the probability of the search finding the best peak, and not just a decent one.

Simulated annealing is used in a wide variety of learning and searching algorithms.

Fitness Scaling Another way to tune the GA is to fiddle with the fitness score. Using a raw fitness score can cause the search to converge on a solution before it has properly explored the full landscape. Early in the search, it is common for just a few lucky genomes to have a high fitness. These genomes can then dominate the breeding pool, causing the search to focus on just one area of the landscape.

To reduce the influence of the raw score during breeding, the genomes can be ranked in order of fitness and then their ranking order used to control breeding preference (Table 6-4). This significantly levels the playing field between the best and worst agents, and encourages the GA to explore more territory in the landscape.

The rank-based fitness does not have to be linear, either. It can follow an exponential, or other, curve. This curve can also adapt as the search progresses, allowing the best fitness scores to become better with time, not unlike simulated annealing.

The simulated annealing temperature control can be applied directly to the fitness scaling problem. Based on the Boltzmann distribution, the scaled fitness can be calculated as:

$$F' = \frac{e^{F/T}}{e^{\mu/T}} \tag{6-21}$$

where:

μ is the mean, or simple average, of all the fitness scores

T is the temperature.

At higher temperatures the fitness values are leveled out, and at lower temperatures the differences in fitness are exaggerated. Table 6-4 shows the Boltzmann scaled fitness at temperatures of 10, 5, and 1.

Finally, the scaling can adapt as the statistical characteristics of the population change. This sigma scaling can accentuate the differences between fitness scores, even when those scores grow closer together as the population converges:

$$F' = \frac{F - \mu}{2\sigma} \quad \textbf{6-22}$$

where σ is the deviation of the fitness population. Both μ and σ are discussed in Chapter 1.

Selection

Once each entity in the population has been assigned some type of fitness value, a number of them are selected for breeding to create mutated child entities.

Table 6-4

Fitness by rank

Agent	Fitness
1	0.34
2	12.97
3	1.46
4	0.01
5	6.28

Agent	Old fitness	Ranked fitness	Fitness (*T*)			Sigma fitness
			10	5	1	
2	12.97	5	2.40	5.76	6361.500	0.398
5	6.28	4	1.23	1.51	7.91	0.094
3	1.46	3	0.76	0.58	0.064	−0.125
1	0.34	2	0.68	0.46	0.021	−0.176
4	0.01	1	0.66	0.43	0.015	−0.191

Table 6-5

Roulette wheel selection

Agent	Fitness $T = 5$	P	R_P
2	5.76	0.659	0.659
5	1.51	0.173	0.832
3	0.58	0.066	0.898
1	0.46	0.053	0.951
4	0.43	0.049	1.000

The new population can be created from the old population several different ways. All of the new population may be children of entities in the old population or some of the old population may be copied. Copied entities may be selected separately from the breeding selection, or they may be limited to the breeding parents.

One way to select the entities that will breed is through tournament selection. This is a naturally inspired selection method that is both simple and easy to implement. To select a parent, a pool of two or more entities is randomly selected from the old population, without any regard to their fitness. From this pool, the entity with the highest fitness is chosen to be a parent.

Another common technique chooses the parents randomly from the population, sometimes call roulette wheel selection. The probability of an entity being selected is proportional to its fitness score. The probability of an entity being selected is:

$$P_i = \frac{F_i}{\sum_j F_j} \qquad \textbf{6-23}$$

Table 6-5 shows the probabilities for our example entities. The last column of the table, R_P, indicates the random number range for any given agent. For example, to select agent 3, the random number must be greater than 0.832 and less than or equal to 0.898.

The mating pool is a combination of roulette wheel selection and tournament selection. A sub-population is selected using a proportional selection system. Then pairs are chosen at random within this mating pool for breeding.

Finally, elitist selection takes your favorite selection and breeding method and copies the entity or entities with the highest fitness into the new population. The elite can be anywhere from one entity to the top 20% or so.

CHAPTER 7

Thinking Logically

This chapter explores the basic concepts of logic and how they apply to proofs and production systems. Then we create a unification library, which is the core component of several logical operations. We apply this unification to two types of logical operation, forward and backward chaining.

For more in-depth information on logic-based AI, look in the bibliography for these books: Russell (1995), Nilsson (1998), Barr (1981), Cohen (1982), Bigus (2001), Michalski (1998), Witten (2000), and Tracy (1996).

Logic

Classic AI systems operate in the realm of logic, discrete searches, game playing, and other symbolic systems. It was thought at one time that logic was the height of human thought and that, with the right representation and construction rules, an intelligence could be created.

Systems of logic are useful for many computer problems. They provide a structure to represent knowledge and methods to process that knowledge.

There are different types of logical systems, several of which are introduced here.

Zero Order

Zero-order logic is also known as propositional logic or propositional calculus. It is essentially a structured version of Boolean logic.

A sentence in propositional logic consists of symbols that are combined using logical operators. Each symbol represents a proposition or sentence of fact. The sentence as a whole then evaluates to *true* or *false*.

An atomic sentence is the truth value *true* or *false*, or a single symbol.

The logical operators are named in Table 7-1, along with some alternative symbols found in other references.

The *and*, *or*, and *not* operators should be familiar by now. Symbols joined by "$\vee$" are also known as a conjunction. Symbols joined by "$\wedge$" are a disjunction. The equivalence operator "$\equiv$" states that the two values or sub-sentences have the same meaning.

Implication is like an *if* sentence. $A \rightarrow B$ reads "If A is *true* then B must also be *true*". The implication operator, however, is odd. If the antecedent

Table 7-1
Logical operators

Operator	Meaning	Alternate
$\wedge$	Logical "and"	&
$\vee$	Logical "or"	|
$\neg$	Logical "not"	~
$\equiv$	Logical equivalence	$\Leftrightarrow$
$\rightarrow$	Implication, "then"	$\Rightarrow$, $\supset$

Table 7-2
Operator truth table

X	Y	$X \wedge Y$	$X \vee Y$	$X \equiv Y$	$X \rightarrow Y$	$\neg X$
F	F	F	F	T	T	T
F	T	F	T	F	T	T
T	F	F	T	F	F	F
T	T	T	T	T	T	F

A is *false*, then the implication as a whole is *true* whether consequence *B* is *true* or *false*.

The functions of these operators are defined by the truth table in Table 7-2.

For example, let us represent the sentence "if the animal is a bird and it cannot fly then it must be an emu or a penguin":

$$(\text{isaBird} \wedge \neg\text{canFly}) \rightarrow (\text{isaEmu} \vee \text{isaPenguin})$$

In even more abstract form, this would be

$$(W \wedge \neg X) \rightarrow (Y \vee Z)$$

where the symbol *isaBird* or *W* represents the proposition "is a bird". The negated proposition, ¬*canFly* is stating the negative fact that it cannot fly.

Given a sentence, it is possible to apply transformations to it, using the rules of inference, to prove that two initially different sentences have the same meaning (or not). For example, the inference rule of Modus Ponens defines this transform:

$$(X \wedge (X \rightarrow Y)) \rightarrow Y$$

When you find the relationship $(X \wedge (X \rightarrow Y))$ in a sentence, you can safely replace it with the simpler form "Y" without changing the function or meaning of the sentence. Though inference rules can be written as implications, they are more often illustrated in this form:

$$\frac{\begin{array}{l} X \rightarrow Y \\ X \end{array}}{Y}$$

The sentences above the line are the premises of the inference rule. The sentence below the line is the conclusion. The meaning is the same as the implication form, but this presentation makes it clear that this is an inference rule.

For each type of operator, there is an inference rule that either adds the operator to a sentence or removes it. The Modus Ponens rule is an *implication elimination* rule. Some inference rules are shown in Table 7-3.

There are also a number of useful equivalences, or tautologies, where two different sentences have the same meaning. Many of the equivalences also have a dual, where the $\wedge$ and $\vee$ operators and the *true* and *false* constants are swapped (Tracy, 1996). Some of these are shown in Table 7-4.

A tautology is a statement that is true under any possible interpretation of its arguments. On the other hand, a fallacy is a sentence that is never true under any possible interpretation. In between lie sentences that are satisfiable, where they are true in at least one interpretation of its arguments.

These rules and equivalences let you manipulate logical sentences, in much same way as the familiar rules of mathematics let you manipulate the equation $3x = 12$ to learn that $x = 4$.

All of these rules and equivalences are used to transform one propositional sentence into another. If a new sentence can be shown to be the same as existing sentences, or combinations of sentences, in the knowledge base, that new sentence is considered to be proven. There are many different ways to perform logical proofs, and entire textbooks are devoted to the subject. Logical proofs are much like algebraic proofs, and proceed by the substitution and inference rules listed here.

A mathematically equivalent version of propositional logic is attributional logic, or variable valued logic. The distinction is mostly one of notation. The symbols in a propositional sentence all represent statements of fact, and resolve to either *true* or *false*. In an attributional sentence, we test attributes of an object against different values, such as a car's size being huge, big, medium, or small.

Table 7-3

Inference rules

Rule	
$\dfrac{X \rightarrow Y \quad X}{Y}$	Modus Ponens, implication elimination
$\dfrac{X \rightarrow Y \quad \neg Y}{\neg X}$	Modus Tolens
$\dfrac{X \quad Y}{X \wedge Y}$	And introduction
$\dfrac{X \wedge Y}{X, Y}$	And elimination
$\dfrac{X}{X \vee Y}$	Or introduction
$\dfrac{\neg\neg X}{X}$	Double negation elimination
$\dfrac{X \vee Y \quad \neg Y}{X}$	Unit Resolution
$\dfrac{X \vee Y \quad \neg Y \vee Z}{X \vee Z}$	Resolution
$\dfrac{\neg X \rightarrow Y \quad Y \rightarrow Z}{\neg X \rightarrow Z}$	Resolution, alternate form

Table 7-4 Equivalences

Theorem	Dual
$\neg(\neg X) \equiv X$	
$X \vee \neg X \equiv$ *true*	$X \wedge \neg X \equiv$ *false*
$X \vee Y \equiv Y \vee X$	$X \wedge Y \equiv Y \wedge X$
$X \vee (Y \vee Z) \equiv (X \vee Y) \vee Z$	$X \wedge (Y \wedge Z) \equiv (X \wedge Y) \wedge Z$
$\neg(X \vee Y) \equiv \neg X \wedge \neg Y$	$\neg(X \wedge Y) \equiv \neg X \vee \neg Y$
$X \vee (Y \wedge Z) \equiv (X \vee Y) \wedge (X \vee Z)$	$X \wedge (Y \vee Z) \equiv (X \wedge Y) \vee (X \wedge Z)$
$X \vee (X \wedge Y) \equiv X$	$X \wedge (X \vee Y) \equiv X$
$X \vee Y \equiv \neg(\neg X \wedge \neg Y)$	$X \wedge Y \equiv \neg(\neg X \vee \neg Y)$
$X \rightarrow Y \equiv \neg X \vee Y$	
$\neg X \rightarrow Y \equiv X \vee Y$	
$X \rightarrow Y \equiv \neg Y \rightarrow \neg X$	
$(X = Y) \equiv (X \rightarrow Y) \wedge (X \rightarrow Y)$	
$(X = Y) \equiv (X \wedge Y) \vee (\neg X \wedge \neg Y)$	

$$(\text{car.size} = \text{big}) \vee ((\text{car.size} = \text{medium}) \wedge (\text{car.price} = \text{expensive})) \rightarrow \text{yuppy\_car}$$

First Order

First-order logic is also known as predicate logic or predicate calculus. It is an extension of zero-order attributional logic that allows you to reason about objects and the relationships between them.

Predicate logic requires a new notation to do this.

In propositional logic, sentences consisted of logical operations on symbolic propositions. In attributional logic, this is extended to tests on the attributes of objects.

Predicate logic performs its tests with function calls that return *true* or *false,* known as predicates. For example, to state that if an object is a car then it has wheels:

$$\text{IsCar}(x) \rightarrow \text{HasWheels}(x)$$

Relationships between objects are tested using predicates with two or more parameters, or arguments. If you feel that car manufacturer y makes terrible cars, you could say that if a car is made by that manufacturer then it is junk:

$$\text{MadeBy}(x, y) \rightarrow \text{Junk}(x)$$

The arguments x and y are variables that stand in for a particular vehicle and manufacturer. Different instances of x and y can be supplied to get different results.

Predicate logic also allows functions to return an object rather than a truth value. For example, to get the manufacturer of a specific car, Firebird:

$$mfg\_of(\text{Firebird})$$

We have introduced four notational elements here: variables in lower case, constants in mixed case, predicates which are in mixed case and have one or more arguments, and functions which are lower case with one or more arguments.

A term is defined as a constant or a function. Note that the "function" label is not strictly correct; it is not a subroutine call, but a way to access information about the referenced object.

An atomic sentence is defined as a single predicate stating a fact. A complex sentence is composed of several terms related through logical operators.

In addition to predicates and functions, we can add the three operators listed in Table 7-5.

The equals operator returns *true* if two objects are indistinguishable under all known predicates and functions. For example, $X = Y$ if for all predicates P, $P(X) \equiv P(Y)$ and for all functions F if $F(X) \equiv F(Y)$.

The other two operators are quantifiers. They introduce the ability to operate on one or more objects from the current domain. If we are reasoning about cars, the domain is the set of all cars plus the set of all car manufacturers, the car colors, upholstery options, price, etc. Whatever is in the knowledge base for the problem at hand is that problem's domain.

The universal quantifier "∀" indicates that a predicate or sentence holds true for all instances in the domain. To say that all cars have wheels:

$$\forall x \ \text{IsCar}(x) \rightarrow \text{HasWheels}(x)$$

Table 7-5

More locial operators

Operator	Meaning
=	Equals
∀	Universal quantifier, "for all"
∃	Existential quantifier, "there exists"

Likewise the existential quantifier "∃" indicates that a predicate or sentence is true for at least one instance in the domain. For example, there exists at least one car that is both red and cheap:

$$\exists x \ \text{IsCar}(x) \wedge \text{Red}(x) \wedge \text{Cheap}(x)$$

If we were operating in a domain composed of only cars, the IsCar(x) would be unnecessary.

The variable x is just a placeholder. Before a variable is used in a statement, it must be attached to a quantifier that defines its use. Universal and existential quantifiers can be mixed to good effect. Order is also important. To use the example from Russel (1995), to say that everybody loves somebody:

$$\forall x \ \exists y \ \text{Loves}(x,y)$$

Using correct parenthesis to demonstrate the scope, this is

$$\forall x(\exists y \ \text{Loves}(x,y))$$

which is quite different from saying that there is someone who is loved by everybody:

$$\exists y \ \forall x \ \text{Loves}(x,y)$$

$$\exists y(\forall x \ \text{Loves}(x,y))$$

The inference rules and equivalences from Tables 7-3 and 7-4 still apply in predicate logic, except using predicates instead of propositions.

With the ∀ and ∃ operators we add some more equivalences, shown in Table 7-6.

Table 7-6

More equivalences

Theorem
$\forall \times P \equiv \neg\exists \times \neg P$
$\forall \times \neg P \equiv \exists \times \neg P$
$\neg\forall \times \neg P \equiv \exists \times P$
$\neg\forall \times \neg P \equiv \exists \times P$

There are also several new inference rules to manage the quantifiers. These rules are more complicated than the propositional inference rules.

The *universal elimination* rule is

$$\frac{\forall x P}{[x / G_x](P)}$$

This introduces a new concept, that of substitution. In this inference rule, P stands for any valid sentence that, most likely, includes the variable x. $[x/G_x\ ]$ indicates that each occurrence of the variable x in P is replaced by G_x.

G_x is a ground term of x. A ground term is a term that has no variables. It is either a constant or a predicate whose parameters, if it has any, are also ground terms.

Given the sentence $\forall x$ *Likes (x, IceCream)* we can substitute *Ben* for the variable X and derive the sentence *Likes(Ben, IceCream)* (Russel, 1995).

The *existential elimination* rule is:

$$\frac{\exists x P}{[x / C](P)}$$

where C is a new constant that does not appear in the current knowledge base. It is getting its definition here, as a side effect of this rule. For example, given $\exists x$ *Kill(x, Victim)* we can use the substitution $[x / Murderer]$ and derive *Kill(Murderer, Victim)*, where *Murderer* is a value that is not currently in the knowledge base.

The *existential introduction* rule is:

$$\frac{P}{\exists x [G / x](P)}$$

where G is a ground term that occurs in P, which is replaced by x which is a variable that does not otherwise occur in the sentence. For example, *Likes(Jerry, IceCream)* transforms to $\exists x$ *Likes(x, IceCream)*.

Second Order

Where first-order logic added the ability to treat the arguments in a sentence as variables, second-order logic treats the predicates themselves as variables. For example:

$$\exists x \ q(x) \wedge r(x) \wedge s(x)$$

This translates into a search for transportation given the substitution:

$$[q/\text{IsCar}, r/\text{Red}, s/\text{Cheap}]$$

$$\exists x \ \text{IsCar}(x) \wedge \text{Red}(x) \wedge \text{Cheap}(x)$$

Of course, different substitutions create different sentences. Second-order logic is useful for reasoning about entire categories of sentences rather than specific sentences. It is also more complicated than first-order logic, and is not normally used in AI.

Using Logic

Logical sentences provide a way to represent knowledge about things and their relationships. Standing by itself, a predicate sentence can give you a small amount of information. Plug in a specific car for x in this sentence:

$$\forall x \ \text{IsCar}(x) \rightarrow \text{HasWheels}(x)$$

And you discover that, tada! It has wheels, too, just like all the other cars.

Though more complicated sentences might generate results that are more interesting, the real fun comes when you combine sets of sentences to discover new facts and relationships.

The process of discovering new information from existing knowledge is called inference. Inference comes in several varieties.

Another way of thinking about logical sentences is as rules. A rule may not only lead to a new fact, but also it could lead to an action. Reasoning with rules parallels the process of reasoning with logical sentences.

All of these logical explorations are performed using first-order predicate logic.

Inference

The simplest form of inference is deductive inference. Deductive inference uses existing sentences, which are facts in a knowledge base, to discover new facts.

For example, given the sentences:

$$\forall y \ \text{IsCar}(y) \rightarrow \text{HasWheels}(y)$$

$$\forall y \ \text{HasWheels}(y) \rightarrow \text{Rolls}(y)$$

and

$$\text{IsCar(Thunderbird)}$$

We can deduce that Thunderbird has wheels. We can take another step and say that it can roll. Deductive reasoning is like following an if–then decision chain.

A more powerful form of inference, inductive inference, learns new rules from specific examples. Developing general rules from examples is the basis of symbolic machine learning. Data mining applies induction rules to discover the relationships and rules implied by a set of discrete facts.

Abductive inference works backwards, taking a hypothesis or sentence which is an observed event or fact. Proving this sentence, using the logical sentences in the system's knowledge base, provides a set of assumptions or causes that lead to this event.

Where induction tries to find general effects given a set of example events, abduction takes a result and tries to infer specific causes, finding explanations for the result.

In all its forms, inference proceeds by stating a hypothesis in the form of a sentence. This sentence is then proven or disproven in the context of existing knowledge. The proof uses the sentences in the knowledge base, transforming them with the various tautologies and inference rules until a sentence can be developed that matches the hypothesis.

Proofs

Proof is central to logical analysis, and Modus Ponens is central to the proof process presented in this section.

Given these sentences in the knowledge base:

1. $\forall x, y \ \text{Hare}(x) \wedge \text{Tortoise}(y) \rightarrow \text{Faster}(x, y)$
2. Hare(Henry)
3. Tortoise(Tim)

Can we prove that Henry is faster than Tim? Note the improved notation in sentence 1, replacing "$\forall x \ \ \forall y$" with the more compact "$\forall x, y$".

$$? \ \text{Faster(Henry, Tim)}$$

First we can combine sentences 2 and 3 with *and introduction*:

4. Hare(Henry) $\wedge$ Tortoise(Tim)

We can eliminate the universal quantifiers in sentence 1 by cleverly picking some constants for x and y.

5. Hare(Henry) $\wedge$ Tortoise(Tim) $\rightarrow$ Faster(Henry, Tim)

Now Modus Ponens gives us our result from 4 and 5:

5. Hare(Henry) $\wedge$ Tortoise(Tim) $\rightarrow$ Faster(Henry, Tim)

4. Hare(Henry) $\wedge$ Tortoise(Tim)

---

6. Faster(Henry, Tim)

So, yes, Henry is faster than Tim.

Now let us introduce the concept of snails and our friend Steve, forgetting about Tim for the moment.

1. $\forall x, y$ Hare(x) $\wedge$ Tortoise(y) $\rightarrow$ Faster(x, y)
2. $\forall x, y$ Tortoise(x) $\wedge$ Snail(y) $\rightarrow$ Faster(x, y)
3. $\forall x, y, z$ Faster(x, y) $\wedge$ Faster(y, z) $\rightarrow$ Faster(x, z)
4. Hare(Henry)
5. Snail(Steve)
6. $\exists x$ Tortoise(x)

Can we prove that Henry is faster than Steve?

? Faster(Henry, Steve)

We assumed in sentence 6 that there must be some Tortoise out there, somewhere. Let us use *existential elimination* to nail it down, even though the constant T does not really mean anything to us:

7. Tortoise(T)

Then we can move through the steps of *and introduction* on 4 and 7, *universal elimination* on 1, and Modus Ponens on that entire mess:

8. Hare(Henry) $\wedge$ Tortoise(T)
9. Hare(Henry) $\wedge$ Tortoise(T) $\rightarrow$ Faster(Henry, T)
10. Faster(Henry, T)

Repeat with Steve:

11. Tortoise(T) ∧ Snail(Steve)
12. Tortoise(T) ∧ Snail(Steve) → Faster(T, Steve)
13. Faster(T, Steve)

Combining the results of 10 and 13 with sentence 3 gives:

14. Faster(Henry, T) ∧ Faster(T, Steve)
15. Faster(Henry, T) ∧ Faster(T, Steve) → Faster(Henry, Steve)
16. Faster(Henry, Steve)

The proof cycles through the steps of *and introduction*, *universal elimination*, and Modus Ponens several times to reach the ultimate conclusion.

This is typical of this type of proof, so is there not a rule that can compress the cycle a bit?

I am glad you asked, since the generalized Modus Ponens does exactly that:

$$\frac{P_1', P_2', \ldots, P_n', (P_1 \wedge P_2 \wedge \ldots \wedge P_n \rightarrow Q)}{\sigma Q}$$

We listed the different premises on one comma-delimited line to keep the rule from growing too tall.

This is complicated, so bear with me through the next example.

There are $n + 1$ sentences in the premise. The first n are the atomic sentences that we are feeding into the rule, some or all of which may be ground terms. The last premise is the implication. This last implication has as many variables as there are atomic sentences.

The conclusion of the rule is Q, with a substitution σ applied to it, so that $(\sigma P_i) = (\sigma P_i')$. What this means is that we replace all of the variables in Q (and which are also in the implication terms T_i) with the appropriate terms from P_i'.

The premise is general, so we need to make it specific using the information from the atomic sentences. This joining is done by substitution, which is represented as Q factored by σ. σ contains the list of substitutions to make to bring the various pieces together.

Restating the snail race, this compresses the proof to:

1. ∀x, y Hare(x) ∧ Tortoise(y) → Faster(x, y)
2. ∀x, y Tortoise(x) ∧ Snail(y) → Faster(x, y)

3. $\forall x, y, z$ Faster$(x, y) \wedge$ Faster$(y, z) \rightarrow$ Faster(x, z)
4. Hare(Henry)
5. Snail(Steve)
6. $\exists x$ Tortoise(x)

We still have to invent the Tortoise T:

7. Tortoise(T)

But now we can jump right to using the generalized Modus Ponens to collect 4, 7, and 1 to give us the conclusion:

Hare(Henry), Tortoise(T), ($\forall x, y$ Hare$(x) \wedge$ Tortoise$(y) \rightarrow$ Faster(x, y))

$\sigma = [x/\text{Henry}, y/T]$

8. Faster(Henry, T)

The substitution transformation σ spreads the constants Henry and T throughout the implication so that it works. This is explained more in the Unification section.

The sentences 7, 5, and 2 likewise collapse in one step to:

9. Faster(T, Steve)

And finally, 8, 9, and 3 give:

10. Faster(Henry, Steve)

Unification

The key to the generalized Modus Ponens above was the ability to find a substitution σ that makes the implication match the other premise terms. The general act of applying a substitution to two terms to make them look the same is unification. The method that finds σ is called Unify, where:

$$\sigma = \text{Unify}(P, Q), \text{ so that } \sigma P = \sigma Q$$

σ is called the unifier of P and Q, and will consist of one or more substitutions for a successful unification, or nothing in the case of failure.

This unification process was going on behind the scenes during the universal eliminations above. While Modus Ponens was central to the proofs, unification is central to Modus Ponens.

Let us look at unification up close, with this knowledge base:

1. Knows(Henry,x)→Hates(Henry,x)
2. Knows(Henry, Steve)
3. $\forall x$ Knows(x, Tim)
4. $\forall x$ Knows(x, Mother(x))

Henry hates everyone he knows, Henry knows Steve, everyone knows Tim, and everyone knows their own mother. Some unifications from this knowledge base are:

1,2: Unify(Knows(Henry,x), Knows(Henry, Steve)) = [x/Steve]

1,3: Unify(Knows(Henry,x_1), Knows(x_3, Tim)) = [x_1/Tim, x_3/Henry]

1,4: Unify(Knows(Henry,x_1), Knows(x_4, Mother(x_4))) = [x_4/Henry, x_1/Henry)]

We are leaving off the universal quantifier. It can be implied for variables when we know that there are no existential quantifiers. Later we will be converting all of our statements into a normalized form, but for now let us just assume it is okay.

The substitution to make sentences 1 and 2 the same is [x/Steve], meaning x is replaced with Steve wherever x occurs.

The next substitution is tricky. Sentences 1 and 3 both use the variable x. Since the variables are just placeholders, and the x in sentences 1 and 3 refers to different variables, we need to rename the variables when we bring the sentences together.

Where there is one substitution that will unify two sentences, there may also be an infinite number of substitutions that will do the same thing. The Unify( ) method tries to return the most general substitution, that is, the substitution that pins the answer down the least.

We will define the algorithm behind unification after we have learned about the next inference process, resolution.

Resolution

While the generalized Modus Ponens proof process provides us with correct proofs, it cannot deal with all forms of sentences.

Modus Ponens only works with a specific subset of logical sentences, known as Horn sentences, named after Alfred Horn who investigated their uses in the early 1950s.

A Horn sentence has the form:

$$P_1 \wedge P_2 \wedge \ldots \wedge P_n \rightarrow Q$$

While this is obviously the form of implication used by Modus Ponens, it less obviously gives us atomic sentences like:

$$true \rightarrow \mathrm{Q}$$

which is the same as just Q.

In the form:

$$P_1 \wedge P_2 \wedge \ldots \wedge P_n \rightarrow false$$

We express the notion:

$$\neg P_1 \vee \neg P_2 \vee \ldots \vee \neg P_n$$

Unfortunately, given the knowledge base:

1. $\forall x_1\, P(x_1) \rightarrow Q(x_1)$
2. $\forall x_2\, \neg P(x_2) \rightarrow R(x_2)$
3. $\forall x_3\, Q(x_3) \rightarrow S(x_3)$
4. $\forall x_4\, R(x_4) \rightarrow S(x_4)$

And we want to prove:

$$?\ S(A)$$

It all falls down. Looking at sentences 1 and 2, it is clear that either $Q(x)$ or $S(x)$ must be true, since either $P(x)$ or $\neg P(x)$ is true. It follows that, no matter what x is, $S(x)$ will then be true.

However, sentence 2 does not convert into Horn clause form, so we cannot use Modus Ponens to drive the proof. Modus Ponens, while sound, is not complete. There are valid inferences that it just cannot make.

Resolution, however, is a complete inference procedure. If it is true, resolution can discover it. Resolution uses the resolution inference rules at the bottom of Table 7-3 instead of Modus Ponens.

Note that resolution combines implications to derive another implication:

$$\frac{\neg X \rightarrow Y \qquad Y \rightarrow Z}{\neg X \rightarrow Z}$$

This also makes it more powerful than Modus Ponens, which only provides atomic sentences for its conclusions.

Interestingly, resolution is a generalization of Modus Ponens. The Modus Ponens implication rule is, of course:

$$\frac{X \qquad X \rightarrow Y}{Y}$$

The term X could also be stated as $false \vee X$ which maps to $\neg\, false \rightarrow X$. This gives:

$$\frac{true \rightarrow X \qquad X \rightarrow Y}{true \rightarrow Y}$$

which is a form of the resolution rule.

We will be working with the conjunctive form of resolution:

$$\frac{X \vee Y \qquad \neg Y \vee Z}{X \vee Z}$$

This inference follows the same logic, if not the same *meaning*, as this mathematical equation:

$$\frac{(X+Y) \qquad +(-Y+Z)}{X+Z}$$

Before we can start applying resolution in proofs, we need to look at how to state our sentences in a manner that resolution recognizes. These are the normalized forms for resolution.

Normalizing Sentences

We can present our knowledge base in either of two ways for resolution. The conjunctive normal form presents the sentences as conjunctions. Those four troublesome sentences in conjunctive normal form are:

1a. $\neg P(x_1) \vee Q(x_1)$

2a. $P(x_2) \vee R(x_2)$

3a. $\neg Q(x_3) \vee S(x_3)$

4a. $\neg R(x_4) \vee S(x_4)$

Conversely, the implicative normal form represents all of the sentences as implications. While conjunctive normal form is historically used more often, all sentences can be freely converted between the two forms. Our four sentences in implicative normal form are

1b. $P(x_1) \rightarrow Q(x_1)$

2b. $true \rightarrow P(x_2) \vee R(x_2)$

3b. $Q(x_3) \rightarrow S(x_3)$

4b. $R(x_4) \rightarrow S(x_4)$

Any first-order logic sentence can be converted into one or the other normalized forms.

There are seven steps to convert an arbitrary first-order sentence into conjunctive normal form. Each step preserves the meaning of the sentence and brings it one step closer to normal form.

1. Eliminate all implications using the equivalence rule $X \rightarrow Y \equiv \neg X \vee Y$.
2. Distribute all negations, so that only atomic sentences are negated and not any complex sentences. There are several equivalence and inference rules that support this step:

$$\neg(X \vee Y) \equiv \neg X \wedge \neg Y$$

$$\neg(X \wedge Y) \equiv \neg X \vee \neg Y$$

$$\neg \forall x\, P \equiv \exists x\, \neg P$$

$$\neg \exists x\, P \equiv \forall x\, \neg P$$

$$\neg\neg P \equiv P$$

3. Adjust the variables so that the various quantifiers of a complex sentence are using different names. For example, the sentence $(\forall x\ P(x)) \vee (\exists x\ Q(x))$ could convert into $(\forall x_1\ P(x_1)) \vee (\exists x_2\ Q(x_2))$, so there is no confusion when we move the quantifiers around in the next step.
4. Now that the variables in the sentence have unique names, we can clump all of the quantifiers on the left-hand side. This would convert the sentence in step 3, for example, into $\forall x_1\ \exists x_2\ P(x_1) \vee P(x_2)$.
5. Remove existential quantifiers. This process is called Skolemization, named after the German logician Thoralf Skolem who worked with this process in the 1920s. Some sentences can lose their existential quantifier using the *existential elimination* inference rule, giving $\forall x_1\ P(x_1) \vee P(A)$. However, there are cases where replacing the variable with a constant stand-in does not work. For example, when the existential variable is embedded within a universal quantifier:

$$\forall x_1\ \mathrm{Car}(x_1) \rightarrow \exists x_2\ \mathrm{Engine}(x_2) \wedge \mathrm{Has}(x_1, x_2)$$

Indicating that all cars have some engine. Replacing x_2 with a stand-in constant, however, gives us the incorrect statement that all cars have the same engine:

$$\forall x_1\ \mathrm{Car}(x_1) \rightarrow \mathrm{Engine}(A) \wedge \mathrm{Has}(x_1, A)$$

The trick, then, is to create a function to replace the variable, instead of a constant:

$$\forall x_1\ \mathrm{Car}(x_1) \rightarrow \mathrm{Engine}(F(x_1)) \wedge \mathrm{Has}(x_1,F(x_1))$$

The function is called the Skolem function. Now that the existential quantifiers have been factored out we can also drop the universal quantifier. We can assume that all variables remaining in the sentence are universally quantified.

6. Distribute $\wedge$ out across $\vee$, so that the sentence contains only conjunctions of disjunctions. For example, $(P \wedge Q) \vee R$ becomes $(P \vee R) \wedge (Q \vee R)$.
7. Flatten parenthesis. $(P \wedge Q) \wedge R$ becomes $(P \wedge Q \wedge Q)$, and $(P \vee Q) \vee R$ becomes $(P \vee Q \vee R)$.

And that's it!

If you want your sentences in implicative normal form, you could add one extra step:

8. Convert the disjunctions to implications. Any negative terms are collected on one side of the sentence, with the positive terms on the right. The sentence then converts from $(\neg Q \vee \neg R \vee S \vee T)$ into $(Q \wedge R) \rightarrow (S \vee T)$.

For example, the race sentences:

1. $\forall x, y$ Hare(x) $\wedge$ Tortoise(y) $\rightarrow$ Faster(x, y)
2. $\forall x, y$ Tortoise(x) $\wedge$ Snail(y) $\rightarrow$ Faster(x, y)
3. $\forall x, y, z$ Faster(x, y) $\wedge$ Faster(y, z) $\rightarrow$ Faster(x, z)
4. Hare(Henry)
5. Snail(Steve)
6. $\exists x$ Tortoise(x)

Convert into normalized form fairly easily. The implications are first converted:

1a. $\forall x, y$ $\neg$(Hare(x) $\wedge$ Tortoise(y)) $\vee$ Faster(x, y)
2a. $\forall x, y$ $\neg$(Tortoise(x) $\wedge$ Snail(y)) $\vee$ Faster(x, y)
3a. $\forall x, y, z$ $\neg$(Faster(x, y) $\wedge$ Faster(y, z)) $\vee$ Faster(x, z)

Then the negations are distributed:

1b. $\forall x, y$ ($\neg$Hare(x) $\vee$ $\neg$Tortoise(y)) $\vee$ Faster(x, y)
2b. $\forall x, y$ ($\neg$Tortoise(x) $\vee$ $\neg$Snail(y)) $\vee$ Faster(x, y)
3b. $\forall x, y, z$ ($\neg$Faster(x, y) $\vee$ $\neg$Faster(y, z)) $\vee$ Faster(x, z)

Finally, we Skolemize the existential variable:

6c. Tortoise(T)

Flattening, we get the final knowledge base of

1d. $\neg$Hare(x) $\vee$ $\neg$Tortoise(y) $\vee$ Faster(x, y)
2d. $\neg$Tortoise(x) $\vee$ $\neg$Snail(y) $\vee$ Faster(x, y)
3d. $\neg$Faster(x, y) $\vee$ $\neg$Faster(y, z) $\vee$ Faster(x, z)
4d. Hare(Henry)
5d. Snail(Steve)

6d. Tortoise(T)

In implicative normal form, it is very close to the original:

1e. Hare(x) $\wedge$ Tortoise(y) $\rightarrow$ Faster(x, y)

2e. Tortoise(x) $\wedge$ Snail(y) $\rightarrow$ Faster(x, y)

3e. Faster(x, y) $\wedge$ Faster(y, z) $\rightarrow$ Faster(x, z)

4e. *true* $\rightarrow$ Hare(Henry)

5e. *true* $\rightarrow$ Snail(Steve)

6e. *true* $\rightarrow$ Tortoise(T)

Proof by Resolution

One complete resolution method is proof by contradiction, or refutation. In it, we try to prove P by failing to prove its inverse $\neg P$.

The resolution rule takes two sentences in normal form, unifies them, and combines them, dropping out the term from one sentence that is negated in the other. For example, let us prove the query that was unprovable under Modus Ponens. The knowledge base in conjunctive normal form is

1. $\neg P(x_1) \vee Q(x_1)$
2. $P(x_2) \vee R(x_2)$
3. $\neg Q(x_3) \vee S(x_3)$
4. $\neg R(x_4) \vee S(x_4)$

And we want to prove $S(A)$ through refutation, so we add:

$$? \neg S(A)$$

Sentences 1 and 3 can be unified with $\sigma = [x_1/x_3]$:

$$\begin{array}{c} \neg P(x_3) \vee Q(x_3) \\ \underline{\neg Q(x_3) \vee S(x_3)} \\ 5.\ \neg P(x_3) \vee S(x_3) \end{array}$$

Likewise, sentences 2 and 5 unified by $[x_2/x_3]$ give us:

6. $R(x_3) \vee S(x_3)$

Combined with 4, we get:

7a. $S(x_4)$

which, as we saw above, is the same as:

7b. *false* $\vee S(x_4)$

Combined with our question *false* $\vee \neg S(A)$ with the unifier $[x_4/A]$ it resolves to:

8. *false*

Since $\neg S(A)$ is *false*, then $S(A)$ must be *true*.

Though we are using sentences with only two terms in them, resolution generalizes in the same as Modus Ponens, giving this form:

$$\frac{\begin{array}{c} P_1 \vee \cdots \vee P_a \vee \cdots \vee P_n \\ Q_1 \vee \cdots \vee Q_b \vee \cdots \vee Q_m \end{array}}{\sigma(P_1 \vee \cdots \vee P_{a-1} \vee P_{a+1} \vee \cdots \vee P_n \vee Q_1 \vee \cdots \vee Q_{b-1} \vee Q_{b+1} \vee \cdots \vee Q_m)}$$

where the unifier $\sigma = Unify(P_a, \neg Q_b)$. The terms P_a and $\neg Q_b$ drop out with all of the remaining terms transferring into the resulting conjunction.

As a human performing the proof, we can look at the knowledge base and make reasonable choices about what to unify and the steps needed to get to our conclusion. Computers do not have our human insights, so they have a more tedious approach.

Code: Unifier

Since unification is fundamental to all of the inference processes explored here, we get our feet wet in logic programming by developing a standalone unifier.

The process of unification is not too complicated. To unify two terms, it helps to think of the terms as lists. For example, the term *P(x, Q(x), z)* consists of the items *(P x (Q x) z)* where *(Q x)* is a sub-list that can stand alone or be stepped into as the situation warrants. Each term can be accessed by its index number, where index=0 is the term's name *P*, index 1 is the next entry *x*, index 2 is *(Q x)*, and the third and last entry is *z*.

A unifiable term can be a constant, a variable, or a predicate. If it is a predicate it will have one or more terms as parameters. The name of a predicate is considered a constant value for the purpose of matching. The algorithm to recursively unify two terms P and Q is:

Set α to the empty set of substitutions.
If P is the same as Q, return the empty α, but okay.
If P is a variable:
  If Q includes P as a term, fail
  ... otherwise return $\sigma = [P/Q]$
If Q is a variable:
  If P includes Q as a term, fail
  ... otherwise return $\sigma = [Q/P]$
If either P or Q is a constant value, fail
(At this point, P and Q are both predicates)
If P and Q have a different number of parameters, fail
For each index entry i in P and Q:
  $\delta = \text{Unify}(P_i, Q_i)$
  Perform the substitution δ on P and Q
  Add δ to α
return α

An example of how this works is shown below, using three examples from Nilsson (1998):

1. $P(x)$
2. $P(A)$

Since these are both terms they might be unifiable, so we step through them one item at a time. The predicate names match so they are okay, but x does not match A. x is a variable so we return the substitution $[x/A]$.

A trickier one is:

1. $P(f(x), y, g(y))$
2. $P(f(x), z, g(x))$

We first find a mismatch at $[y/z]$. Applying that substitution we get:

1b. $P(f(x), \mathbf{z}, g(\boldsymbol{z}))$

2b. $P(f(x), z, g(x))$

Moving forward in the sentences, the next mismatch is at the parameter of the *g()* functions, giving the substitution $[z/x]$, giving the final match:

1c. $P(f(x), \boldsymbol{x}, g(\boldsymbol{x}))$

2c. $P(f(x), \boldsymbol{x}, g(x))$

A final example:

1. $P(f(x, g(A, y)), g(A, y))$
2. $P(f(x, z), z)$

The first mismatch is at the *g()* function and *z*. This gives the substitution $[z, g(A, y)]$:

1. $P(f(x, g(A, y)), g(A, y))$
2. $P(f(x,\boldsymbol{g(A, y)}), \boldsymbol{g(A, y)})$

And it is done.

Most of the code for unification is used to represent the terms.

The class diagram for the logic support classes is shown in Figure 7-1.

Term.java

aip.logic.Term A term is a generic class that provides a base for the other logical objects. It provides the basic operations used across its various child classes. A summary of *Term* is given in Table 7-7.

copy() This method creates a duplicate of the *Term*.

equals(Term term) The *equals()* test is used to verify that the contents of two terms are exactly the same.

ground() A ground term is a term that has no variables; it is completely filled in. Certain operations require ground terms. This method only has

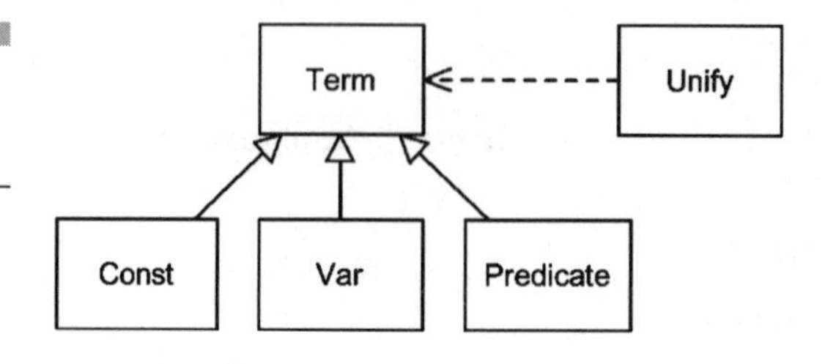

Figure 7-1 Unify class diagram

Table 7-7

Term summary

```
Construction
                         Term (String name)
       abstract Term  copy ()

Operations
    abstract boolean  equals (Term term)
    abstract boolean  ground ()
    abstract boolean  includes (Term term)
        abstract int  size ()
       abstract Term  get (int index)

Access
              String  getName ()
```

to think when the term is a predicate since constants are always ground terms and variables never are.

includes(Term term) The *includes()* method is mostly for predicates, but is included as a generic *Term* method to simplify processing in the higher level objects.

size( ) Using the list analogy, each piece of a *Term* can be retrieved by its index number. For constants and variables, the only legal index is 0. However, predicate parameters allow for more variety.

get(int index) Retrieve part of this term as another term. Predicate names are returned as constants.

Const.java

aip.logic.Const The *Const* term has only its name. All of the other *Term* methods are stubbed out.

Var.java

aip.logic.Var Variables are like constants but with different semantics. Variables are stand-ins for actual values. As such, variables are the only part of a statement that can be replaced during unification.

For most processing, the *Var* is just a name. However, the *Var* does dual duty as the unit of substitution. As such, it also has a value. A summary of *Var* is given in Table 7-8.

Predicate.java

aip.logic.Predicate The predicate is the "word" used in logic sentences, while the constants and variables are the "letters" that make up the

Table 7-8

Var summary

```
Construction
                Var (String name)
        Term    copy()

Operations
     boolean    equals (Term term)
     boolean    ground ()
     boolean    includes (Term term)

Acess
         int    size ()
        Term    get (int index)
        Term    getValue ()
        void    setValue (Term value)

Public Methods
      String    toString ()
```

Table 7-9

Predicate summary

```
Construction
                Predicate (String name, Term param)
        void    add (Term param)
        Term    copy ()

Operations
     boolean    equals (Term term)
     boolean    ground ()
     boolean    includes (Term term)
        void    substitute (int index, Term term)

Acess
         int    size ()
        Term    get (int index)

Public Methods
      String    toString ()
```

predicate word. A summary of *Predicate* is given in Table 7-9. A few methods are unique to *Predicate*.

Predicate(String name, Term param) Predicates should have at least one parameter, so the first parameter is defined in the constructor along with the name. Occasionally, for mechanically constructed predicates, you do not have the first parameter on hand during construction. For these cases, *param* may be *null*.

add(Term param) Add another parameter term to the *Predicate*.

substitute(int index, Term term) This method replaces the *Term* at the given index position with the specified *Term*. Since index 0 is the name, it may not be substituted.

Unify.java

aip.logic.unify The *Unify* class is a utility class with static methods. These methods either find or apply a list of substitutions. Though we only use the basic *Unify(Term P, Term Q)* method in this example, we will discuss all of the methods in *Unify* now. These will make more sense once you have explored the later code examples.

A summary of *Unify* is given in Table 7-10.

unify(Term P, Term Q) This is the basic unification method, which finds the substitution that makes *P* and *Q* the same. While it returns a list of *Var* objects that describe the substitution, it also applies this substitution internally. Upon a successful unification *P* and *Q* will be identical.

Before *Unify()*:

$$P = \text{“}P(f(x), y, g(y))\text{”}$$

$$Q = \text{“}P(f(x), z, g(x))\text{”}$$

After *Unify()*:
Substitution = $[y/z, z/x]$

$$P = \text{“}P(f(x), x, g(x))\text{”}$$

$$Q = \text{“}P(f(x), x, g(x))\text{”}$$

unifyBack(Rule rule, Term Q) A rule is a logical statement, sentence, implication… actually a positive Horn clause. This consists of one or more predicates for the conditions, or antecedents, of the rule, and another predicate that is the result, or consequence.

Table 7-10

Unify summary

Unification

```
    ArrayList  unify (Term P, Term Q)
    ArrayList  unifyBack (Rule rule, Term Q)
    ArrayList  unifyForward (Rule rule, Term Q)
```

Substitution

```
         void  substitute (Predicate pred, ArrayList sub_array)
         Term  substitute (Var var, ArrayList sub_array)
```

Back-chaining unification tries to unify a term with the rule's consequence. If it succeeds, that substitution is applied across the entire rule and the substitution list is returned. Note the modified notation, which is explained in a later section.

Before *Unify()*:

$$R = \text{"Faster}(x, z)\text{:=Faster}(x, y) \text{ \& Faster}(y, z)\text{"}$$

$$Q = \text{"Faster(Henry, Steve)"}$$

After *Unify()*:

$$\text{Substitution} = [x/\text{Henry}, z/\text{Steve}]$$

$$R = \text{"Faster(Henry, Steve):=Faster(Henry, }y) \text{ \& Faster}(y, \text{Steve)"}$$

unifyForeward(Rule rule, Term Q) Forward-chaining unification works with the antecedents of the rule. It steps through these one at a time, attempting to unify with Q. As soon as it finds a term that unifies, it applies the substitution to the entire rule.

Before *Unify*:

$$R = \text{"Hare}(x) \text{ \& Tortoise}(y) \rightarrow \text{Faster}(x, y)\text{"}$$

$$Q = \text{"Hare(Henry):}$$

After *Unify:*

$$\text{Substitution} = [x/\text{Henry}]$$

$$R = \text{"Hare(Henry) \& Tortoise}(y) \rightarrow \text{Faster(Henry, }y)\text{"}$$

Unifier.java

aip.app.unifier.Unifier *Unifier* is a simple command-line program that creates predicates, attempts to unify them, and prints the results to the console.

Using Unify The various logical proofs and operations described in this chapter revolve around the unify process. A proof is a lot like a search–you are searching through a knowledge base of rules and fact trying to find the ones that will bring your proof closer to completion.

There are different search strategies available that try to provide completeness at the same time as they minimize the work being done. Descriptions of these can be found in any of the references listed at the beginning of this chapter.

There is a constraint in this implementation of *Unify* that limits its functionality. It does not know about negated terms, which are an important feature for resolution proofs. However, we can use this *Unify* for Modus Ponens-type proofs.

Code: Backward Chaining

Backward chaining provides an efficient proof method for a specific form of logical sentence, the Horn clause:

$$P_1 \wedge P_2 \wedge \ldots \wedge P_n \rightarrow Q$$

Another way to think of the Horn clause is as a rule. The conjunction part of the implication contains the antecedents of the rule. These are the terms that must all be true for the rule to be fired. The Q term is the consequence of firing the rule.

Backward chaining takes a hypothesis, or goal statement, in the form of a Horn clause without a consequence and sees whether this hypothesis is true in the context of the rules and facts in the knowledge base.

Facts are Horn clauses without any antecedents, just a consequence. Rules tend to be stated in terms of predicates with variables. The variables are then grounded using facts from the knowledge base.

To represent Horn clauses in a way that can be typed on your typical keyboard, we need to revisit the subject of notation. The notation we have used so far describes a Horn implication as:

$$P_1 \wedge P_2 \wedge \ldots \wedge P_\mathrm{n} \rightarrow Q$$

For backward chaining, we reverse the order so it is easier to read for this algorithm:

$$Q \leftarrow P_1 \wedge P_2 \wedge \ldots \wedge P_n$$

And then change the implication and "and" symbols to keyboard equivalents:

$$Q \text{ :- } P_1 \ \& \ P_2 \ \& \ \ldots \ \& \ P_n$$

Backward chaining is presented here using a depth-first search. The process begins with a knowledge base of rules and known facts, such as these familiar sentences:

1. Faster(x,y) :- Hare(x) & Tortoise(y)
2. Faster(x,y) :- Tortoise(x) & Snail(y)
3. Faster(x,z) :- Faster(x,y) & Faster(y,z)
4. Hare(Henry)
5. Snail(Steve)
6. Tortoise(T)

Then we ask a question in the form of one or more terms. If there are multiple terms in the question, they must all be made true. Our tests use only a single term as the question, such as:

:- Faster(Henry, Steve)

We look for rules that have a consequence that matches the term to be proved. In this case, there are three possibilities: sentences 1, 2, and 3. Though the program would have to test each of these, we can see that only 3 will prove to be true. Let us unify the question term with sentence 3:

[x/Henry, z/Steve]

? Faster(Henry, Steve) :- Faster(Henry, y) & Faster(y, Steve)

This unified rule is not grounded yet, so it cannot fire. Neither does it fail. To see if this rule is true, we must try to prove all of its antecedents, so we now have two questions to answer:

? Faster(Henry, y)

? Faster(y, Steve)

The first question can be unified with KB sentence 1:

[x/Henry]

? Faster(Henry, y) :- Hare(Henry) & Tortoise(y)

Of course, *this* rule is not grounded either! So we now ask the relevant questions:

? Hare(Henry)

? Tortoise(y)

We discover that *Hare(Henry)* matches fact 4, so that is true. *Tortoise(y)* can unify with fact 6:

[y/T]

? Tortoise(T)

This term is grounded and, hence, true. We can now accept that *Faster(Henry, y)* is true.

Now we back up the tree to try and prove *Faster(y, Steve)*, which we do. Since both *Faster(Henry, y)* and *Faster(y, Steve)* can be resolved to *true*, *Faster(Henry, Steve) :- Faster(Henry, y) & Faster(y, Steve)* is true. So, of course, we know that our original query is *true*.

To implement backward chaining, we add two additional classes to our logic system and use the *unifyBack()* method of *Unify*. The class diagram for *Rule* and *Goal* is shown in Figure 7-2.

Goal keeps track of the predicates that are being proved, while the knowledge base is built up of *Rule* objects.

Rule.java

aip.logic.Rule A rule consists of the antecedent predicates, called the *body*, and a single consequence predicate, the *result*. While, for our current purposes, the result is a *Predicate*, it could also be an action or command that can be used to drive an agent. Not that this would be very

Figure 7-2 Rule and goal class diagram

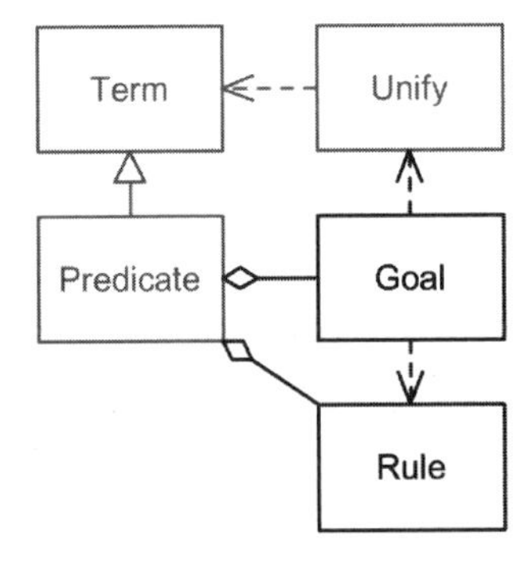

Table 7-11

Rule summary

```
Construction
                Rule ()
          Void  setResult (Predicate term)
          void  add (Predicate term)
          Rule  copy()

Operations
          void  substitute (int index, Predicate term)
       boolean  fire ()

Access
           int  size ()
     Predicate  get (int index)
     Predicate  getResult ()
```

useful during backward chaining, but in the next section on forward chaining an active result could be triggered during the forward-chaining process.

A summary of *Rule* is given in Table 7-11.

setResult(Predicate term)
getResult( ) The result of the rule is the consequence of the rule, and is represented by a *Predicate*.

size( )
add(Predicate term)
get(int idex) The antecedents, or body, of the rule consist of predicates. The *size()* method returns the number of antecedents, and *get()* retrieves them.

substitute(int index, Predicate term) Substitute a predicate in the antecedents with a different one. This is used during unification to mutate the rule into a more grounded form.

fire( ) If all of the antecedents of the rule are ground terms, then *fire()* returns *true*.

Goal.java

aip.logic.Goal A *Goal* holds the predicates that currently need to be proved. *Goal* operates in two modes. In the "or" cycle, it tries to find one rule out of the database that can be used to prove a single goal predicate. In the "and" cycle it must prove all of the predicates in a predicate stack. These cycles alternate. When a rule unifies with a single goal, its

Table 7-12

Goal summary

```
Construction
                Goal (Rule goal)
                Goal (Predicate goal)
          void  setTrace (boolean trace)

Operations
       boolean  prove (ArrayList rule_list)

Public Methods
     ArrayList  getTrace ()
```

antecedents are then pushed into a goal stack so they can all be proved. Each antecedent, then, is popped from the stack and the cycle recurses.

A summary of *Goal* is given in Table 7-12.

Goal(Rule goal)
Goal(Predicate goal) A *Goal* initialized with a *Rule* must prove all of the antecedents to the rule.

A *Goal* initialized with a *Predicate* must prove just that predicate.

prove(ArrayList rule_list) Returns *true* if the proof could be completed.

This method attempts to prove the *Goal* using the list of *Rule* objects in the *rule_list*. The proof is recursive, where each rule that unifies with a predicate creates another *Goal* that tries to prove the antecedents, which are themselves *Goal*s to be proved. This is illustrated in Figure 7-3.

The root goal tries to prove itself using one of the rules in the rule list, or knowledge base. Each rule that unifies with the goal is then expanded so each of its antecedents becomes another goal. All of the antecedents, however, must be proved, not just one. This continues until either the proof tree is fully grounded and true, or a goal fails to unify with any rule and fails.

getTrace() If the proof came back *true*, then *getTrace()* returns a list of the unified *Rule* objects used for the proof.

Resolver.java

aip.app.resolver.Resolver *Resolver* is a simple command-line program that creates a knowledge base of rules and facts, tests a variety of questions against them, and prints the results to the console.

Each variable is defined with a counter, which is the same as the rule number it is associated with. This is to force each rule to have unique

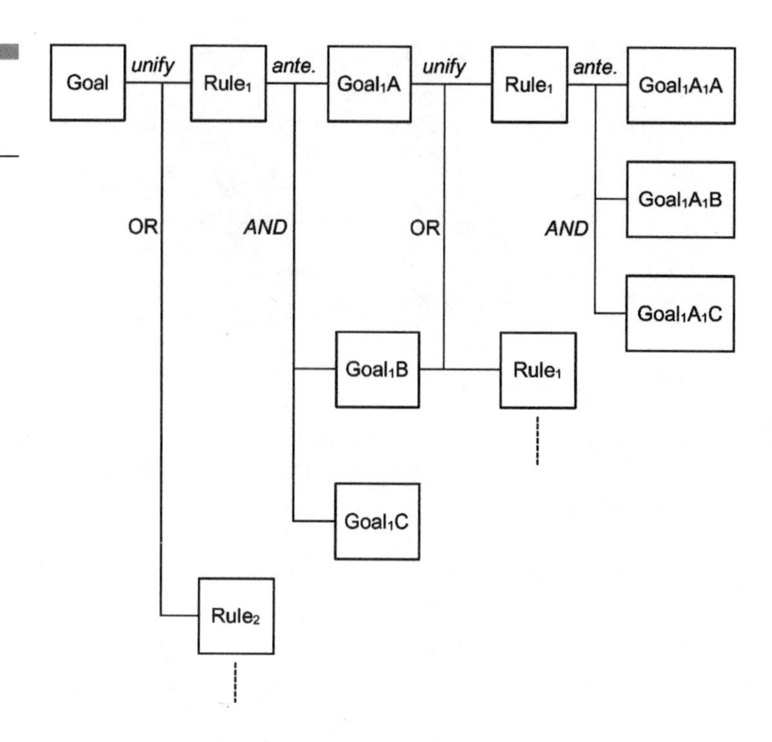

Figure 7-3
Depth first resolution

variables, to avoid confusion during unification. For a system with a rule parser, these variable suffixes should be applied automatically.

Note that there is a *TRACE* constant at the top of the *Resolver* class that you can set to *true* to enable diagnostic trace output.

Code: Forward Chaining

Forward-chaining systems are also known as production systems, since they produce new facts from their knowledge base.

If you extend the *Predicate* class to access a variable system and perform tests, you can open up the system to a whole new range of antecedents. Extending consequences so they can set values in a variable system or trigger actions, the forward-chaining system can be used to control agents in an environment.

Forward chaining is easier to describe using the familiar implication sentence ordering:

1. Hare(x) & Tortoise(y) $\rightarrow$ Faster(x, y)
2. Tortoise(x) & Snail(y) $\rightarrow$ Faster(x, y)
3. Faster(x, y) & Faster(y, z) $\rightarrow$ Faster(x, z)

These make up the rules of the production system. Facts, which are simply grounded predicates, are held separately and are added by the *chain()* method. As each fact is added to the system, it is processed against the rules to see what other facts can be inferred.

To infer new facts, each rule in the knowledge base is examined. If any of its antecedents can be unified with the new fact, that rule is tested against the current set of facts to see if it can be fully grounded. If we manage to ground the rule, its consequence is then added to the list of new facts.

If the consequence is an action or other extension, it may not create a new fact but a new behavior.

Adding the fact *Hare(Henry)* does not cause any rules to ground, though it does match an antecedent in rule 1.

The next fact gets us a bit farther:

$$\text{Tortoise}(T)$$

This unifies with a predicate in rule 1, giving:

$$\text{Hare}(x) \mathbin{\&} \text{Tortoise}(T) \rightarrow \text{Faster}(x, T)$$

Working with the existing list of facts, we can ground the rule:

$$\text{Hare(Henry)} \mathbin{\&} \text{Tortoise}(T) \rightarrow \text{Faster(Henry}, T)$$

And we can add *Faster(Henry, T)* as a new fact. This, of course, may be able to chain against more rules. We recursively check this new fact against the knowledge base.

While it unifies with rule 3, we cannot ground that rule yet, so no more facts are created. A possible run looks like:

```
1. Hare(x1) & Tortoise(y1) → Faster(x1, y1)
2. Tortoise(x2) & Snail(y2) → Faster(x2, y2)
3. Faster(x3, y3) & Faster(y3, z3) → Faster(


Fact: Hare('Henry')
→ Hare('Henry')


Fact: Tortoise('T')
→ Tortoise('T')
→ Faster('Henry', 'T')
```

```
Fact: Snail('Steve')
→ Snail('Steve')
→ Faster('T', 'Steve')

Fact: Tortoise('Tim')
→ Tortoise('Tim')
→ Faster('Henry', 'Tim')
→ Faster('Tim', 'Steve')
```

Note that the actual *Chainer* example has more rules.

To implement forward chaining, we add one more class to the logic system and use the *unifyForward()* method of *Unify*. The class diagram for *Fact* is shown in Figure 7-4.

Fact is a holder for grounded predicates, and provides the recursive chain method needed to generate new facts from the knowledge base.

Fact.java

aip.logic.Fact A *Fact* holds a single grounded predicate. Its *chain()* method searches through the list of rules in the knowledge base looking for a rule that has an antecedent that will unify with this predicate. If it finds such a rule, it then tries to ground, or resolve, the rule against the list of existing facts.

If *chain()* is able to fully ground a rule, its consequence is added to the list of new facts and is, in turn, chained to see if it generates even more facts.

A summary of *Fact* is given in Table 7-13.

chain(ArrayList rule_list, ArrayList fact_list, ArrayList result_list) The *chain()* method takes this fact and adds it to the *result_list*. It then tries to chain it against the antecedents of all the rules in the *rule_list*. Any additional facts that are generated from rules that fire are also added to the *result_list* and are recursively chained against the *rule_list*.

The calling program will need to deal with the *result_list*, and may want to add the results into the main *fact_list* for later reference.

exists(ArrayList fact_list) *exists()* returns *true* if this fact already exists in the *fact_list*.

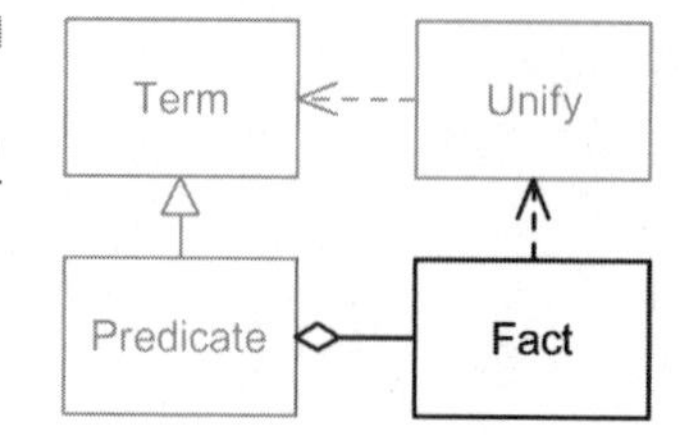

Figure 7-4 Fact class diagram

Table 7-13

Goal summary

Construction

```
              Fact (Predicate fact) throws Exception
```

Operations

```
        void  chain (ArrayList rule_list, ArrayList fact_list,
              ArrayList result_list)
     boolean  exists (ArrayList fact_list)
```

Chainer.java

aip.app.chainer.Chainer *Chainer* is a simple command-line program that creates a knowledge base of rules, adds a variety of facts to them, and prints the results to the console.

Rule 9 in *Chainer*, "Reptile(x) $\rightarrow$ Legs(x, 4)", illustrates a limitation of this simplified *Unify*. It would be better to state rule 9 as *Reptile(x)* $\wedge$ $\neg$*Snake(x)* $\rightarrow$ *Legs(4)*, so we can make a snake exception. However, we need negated predicates to do this. I leave this modification as an exercise for the reader.

Note that there is a *TRACE* constant at the top of the *Chainer* class that you can set to *true* to enable diagnostic trace output.

Knowledge Representation

The first-order predicate logic described in the previous section can be used to describe generic relationships as well as specific instances. In the *Chainer* example, one relationship is *Snail(x)* $\rightarrow$ *Mollusk(x)*, saying that all snails are mollusks, we then say *Snail(Steve)* indicating that *Steve* is an instance of the *Snail* category.

Sometimes a list of rules is not the most intuitive way to visualize a knowledge base.

Other presentations include the semantic network and the frame.

Semantic Network A semantic network is a graph of nodes that are objects, categories, or attributes, which are connected by arrows that represent relationships. The complete *Chainer* knowledge base is shown as a semantic network in Figure 7-5.

Semantic networks are freely convertible to first-order predicate logic.

Frames A similar approach to knowledge representation is the frame-based system. Each frame contains slots for attributes of that frame.

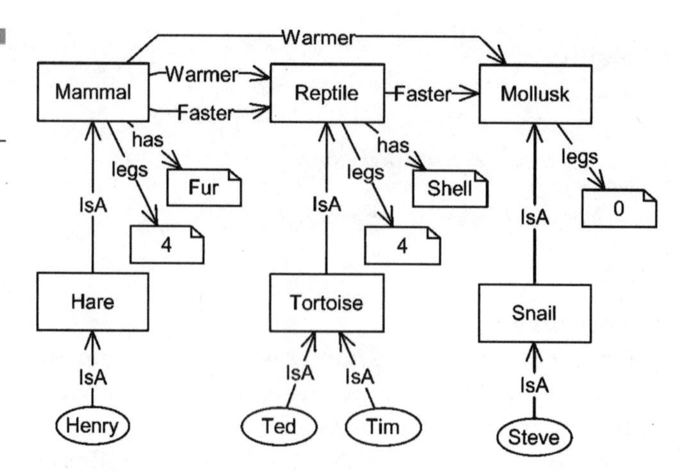

Figure 7-5 Chainer semantic network

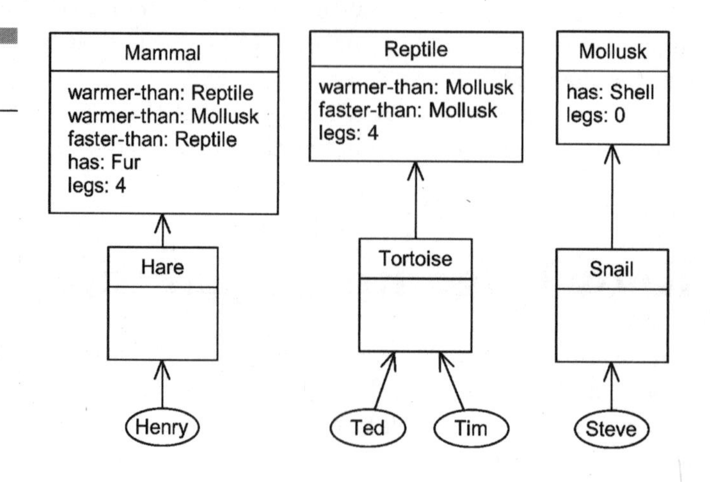

Figure 7-6 Chainer frames

A frame, then, represents an object or class of objects. Frames may be referenced by more general frames.

Frame-based systems are a lot like the objects in object-oriented programming.

The *Chainer* database, in frame form, is shown in Figure 7-6.

Different representations help you to think about the data in different ways. Being able to convert them into FOPL gives you a common set of tools for processing the information.

Probability So far, all of our reasoning has been done with absolute rules. *This set of antecedents leads to that consequence*. But what if the knowledge base is not certain? Or what if there are several possible outcomes for a given fact?

The implications can be assigned probability factors. Under the system defined so far, each implication has a probability of 1.

As reasoning progresses through the tree, the probabilities are factored together. Of course, this means that one fact can trigger several rules, each with different levels of uncertainty, but the forward-chaining algorithm should work fine for that.

If you want to try backward chaining, to discover the probability that a statement is true, the and/or tree structure will have to adapt. Instead of proving just one rule to support a statement, it becomes necessary to explore all non-zero probability rules. This will provide, then, several inference paths to the goal statement, each with a different probability.

Note the similarities to the Markov probability system from Chapter 4, and how the Markov finite-state network maps to the semantic network described earlier.

It all comes down to the same thing, using different tools to process the data.

Probabilistic semantic trees are the domain of Bayesian networks. These are explored further in Tozour (2002), Tracy (1996), and Witten (2000).

Note that instead of probabilities, the implications could have activation levels. These are essentially the same thing as probabilities in a semantic network, but are more relevant to forward chaining than backward-chaining applications. And, of course, the activation levels could be defined using fuzzy logic. The rule antecedents could be fuzzy logic tests, returning activation levels. These could be combined with the other fuzzy antecedents, giving an overall activation level for the consequence.

CHAPTER 8

Supervised Neural Networks

This chapter introduces the concept of simulated intelligence, featuring the connectionist model of computation. In short, artificial neurons and neural networks (NNs). We begin this tour by looking at simulations of individual neurons. We then move on to multiple neurons working together, and then multiple layers of neurons. All these models use supervised learning to train their responses.

Once the basics are in place, we look at data normalization to make the inputs and outputs of the networks more palatable.

The chapter wraps up with an associative memory network model and some ideas for further study.

Chapter 9 continues on the topic of simulated intelligence, looking at NNs using unsupervised learning.

Simulated Intelligence

While the AI techniques explored so far implement symbolic intelligence (the searches and the logical representations), this chapter looks at simulated intelligence, the "natural" computation techniques.

The classic intelligence techniques process symbols. In the first-order predicate logic, the symbols are explicit and easy to see, however, the searches are also symbolic. They operate on fields with values that have "meaning". Natural, or neural, computation uses networks of simple, interconnected processing elements, or neurodes. The network receives external stimulus and this propagates through the network until some kind of result pops out the other end; thought as reflex. Internally, there are no symbols, no explicit representation of knowledge.

In symbolic computation, the knowledge is separate from the processing. The information in symbolic AI is stored in a pile somewhere and a set of rules and algorithms process it to find a result. Learning in these classic systems involves discovering new rules that can be used to process the information.

In neural computation, the knowledge is distributed across the network, implied by the state and structure of each processing element. Processing is done reflexively, filtering input signals against the neural states. Learning is done by adjusting neural states across the network, which has the effect of changing the distributed memory of the network.

While symbolic AI tries to use the rules of logic to define a spark of intelligence, neural AI approaches the problem from the bottom up, simulating what looks like the relevant aspects of biological systems to generate intelligence.

Many classical techniques are statistical in nature, from Bayesian Networks, the Markov techniques, and others. Logical learning is also done statistically, discovering rules from data (Witten, 2000). But at the core, these statistical techniques are operating within a framework of symbols and logical structure.

NNs are also statistical, discovering patterns and regularities in the data and then molding themselves around them in order to generate "right" answers. NNs are especially useful when the data to be processed is imprecise, noisy, or chaotic. Since networks discover the structure in data, they are also useful when the rules are not known.

I like to call the biological computing techniques simulated intelligence, since the original "artificial intelligence" label is closely associated with the symbolic techniques.

Neural Models

NNs are nothing more than collections of simple processing elements, modeled on neurons. Most NNs use stylized neural models that capture only the broadest actions of the biological neuron.

Biological Neurons

A neuron is a complex system of molecules and ions sitting in a bath of even more molecules and ions. The skin of a neural cell is a permeable membrane that allows these molecules and ions to pass in and out.

Like any other processing device, the neuron has inputs, a processing core, and outputs. It also has systems to restore the cell's state over time so that it is always ready to perform its job. Figure 8-1 shows a generic neuron shape. There are many types of neurons, and they often have wildly different shapes.

The dendrites collect chemical signals from other neurons. A byproduct of this chemistry is an electrical charge in the neural cell. When the charge builds up to a threshold level the neuron "spikes", releasing a pulse of energy down the axon. This spike is known as the action potential. When this pulse reaches the axon terminals, the terminals release a puff

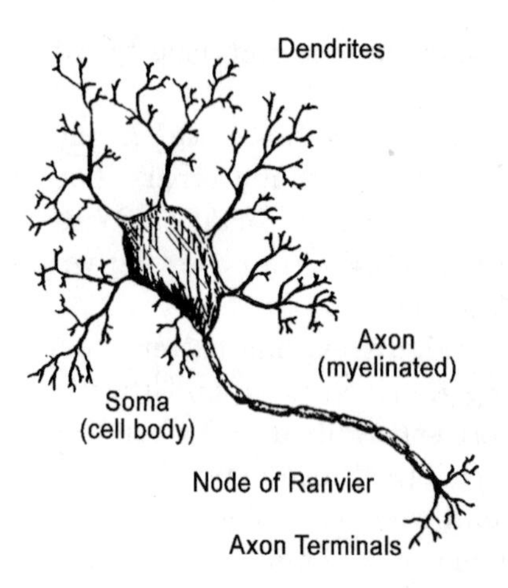

Figure 8-1
Neuron

of chemicals. These chemicals, in turn, cross a tiny gap to be received by the dendrites of other neurons.

There are two types of input to the neuron. At some junctions the signal coming in stimulates, or excites, the neuron. At others the signal reduces, or inhibits, the charge in the neuron. All of the signals are added together over time, until they trigger an action potential.

As the signals accumulate at the dendrites, the various membranes in the neuron are working to keep the cell at its preferred chemical and electrical balance. This causes the input charges to decay over time.

In the "real world" the environment is noisy and lossy. Electrical and chemical signals are buffeted by random noise, charges get lost as they move around, chemicals leak. If the neuron tried to communicate to its neighbors with subtle voltage levels at its axon terminals, the information signal could easily be lost in the background noise. This is why the neuron spikes. An essentially digital, all-or-nothing, pulse is hard to miss since it rises far above the background noise. To signal higher levels of activation, the neuron sends more pulses closer together. If the neuron is just idling, bored, it may send a pulse every once in a while.

The axon on some neurons has to travel a long way through the body. The neural signal, however, fades as it travels. There are two mechanisms built in the axon to fight this signal loss.

There is a protective sheath around some axons, called myelin, that acts as an insulator, keeping the signal strong and longer. There are gaps in the myelin called the nodes of Ranvier where the axon has access to the chemical soup it lives in. At these points the action potential is regenerated.

Once an action potential has been created by the neuron or at any of its nodes of Ranvier, the ability to spike is temporarily lost. This dead time is known as the refractory period. One effect of the refractory period is to force the action potential to travel in one direction, down the axon. When a node of Ranvier creates an action potential, the upstream node has already spiked and cannot do it again, so the signal fades out in that direction. Downstream, however, it reaches a fresh node which then fires. This continues until the signal reaches the axon terminals.

An exception to the action potential model of neural behavior can be found in the sensory nerves. These cells do not spike, but raise and lower the intensity of their output signal as their stimulus changes. An action potential cell is closely joined with the sensory cell, to immediately convert this amplitude-based output into the more robust frequency-based form used by the downstream neurons.

If you are interested, an excellent introduction to neurons and neural systems can be found in *Neurons and Networks* by John Dowling (1992).

The electro-chemical action of a neuron can be thought of as an electrical circuit, and many books present a ladder of resistors and capacitors to describe the neuron's behavior.

There are a number of mathematical models for biological neuron behavior, the first and most famous of which is the Hodgkin–Huxley (HH) model.

Code: Hodgkin–Huxley Neural Model

While the environment of a neural cell is chemical, its operation is essentially electrical. This is because the active components of neural communication are ions, which are electrically charged atoms.

The movement of these ions from one place to another provides electrical current, and an accumulation of ions inside the neural cell gives it an electrical charge relative to its environment. This flow of ions is regulated by tiny portals that regulate the motion in and out of the neuron.

A.L. Hodgkin and A.F. Huxley meticulously studied the giant axon of the squid and were able to quantify its electrical behavior (Hodgkin, 1952).

The HH model uses the sodium ion, Na^+, and potassium ion, K^+, channels. These are the mechanisms in the axon's wall, or membrane, that control ion flow in and out of the cell. Note that the action potential model of the neuron's soma is an extension of the HH axon model, with the addition of calcium ion flow, Ca^+.

Electrically, the HH model of the axon membrane is represented by the schematic in Figure 8-2.

The membrane has a capacitance C_M that is essentially constant. The Na^+ and K^+ channels have a voltage offset, and a variable resistance R_{Na} and R_K through the membrane. The HH model actually deals with the conductance g of the membrane, and not resistance. Conductance is the capacity to conduct the ions, the reciprocal of resistance.

The membrane also has a slight leak, due to various environmental factors, represented by R_L.

The overall current flow for the membrane is the sum of its component currents:

$$I = I_{Na} + I_K + I_L \qquad \textbf{8-1}$$

Any external stimulation, such as from the synapses of other neurons, is added to this.

Each current flow component depends on the voltage potential, V_m, across the membrane, relative to the voltage offset, V_i, for that ion, and the current conductance g_i of that ion's channel:

$$I_i = g_i(V_m - V_i) \qquad \textbf{8-2}$$

The conductance is controlled by two factors. On the one side, there is the rate of ion transfer from outside the cell to the inside, α. Balancing

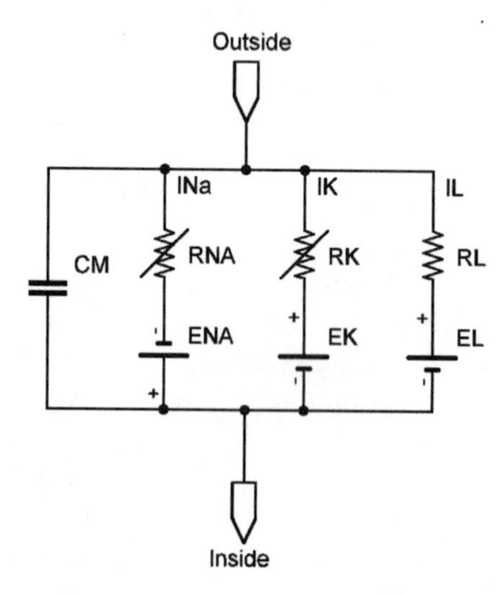

Figure 8-2
Axon membrane schematic

that is the rate of transfer from the inside to the outside, β. These rate factors are not time dependent, but rely only on the membrane's voltage potential.

The conductance of the K^+ channel is controlled by the factor n, which is determined by its flow α into the cell and β out of the cell:

$$\alpha_n = \frac{0.01(V_m + 10)}{\exp\left(\frac{V_m + 10}{10}\right) - 1}$$
$$\beta_n = 0.125 \exp\left(\frac{V}{80}\right) \qquad \textbf{8-3}$$
$$\dot{n} = \alpha_n(1-n) - \beta_n n$$

The K^+ conductance is then:

$$g_K = 36n^4 \qquad \textbf{8-4}$$

The Na^+ conductance is more complex. This rate appears to be determined by a second-order equation. To keep the model simple, this can be reduced to two first-order equations. The model assumes that the Na^+ channels are activated by molecules around the membrane, assuming these channels have not already been de-activated by inhibiting molecules. The activating molecule flow is represented by rate factor m and the inhibiting molecule flow by h. These both have appropriate α and β control factors:

$$\alpha_m = \frac{0.1(V + 25)}{\exp\left(\frac{V + 25}{10}\right) - 1}$$
$$\beta_m = 4 \exp\left(\frac{V}{18}\right) \qquad \textbf{8-5}$$
$$\dot{m} = \alpha_m(1-m) - \beta_m m$$

$$\alpha_h = 0.07 \exp\left(\frac{V}{20}\right)$$
$$\beta_h = \frac{1}{\exp\left(\frac{V + 30}{10}\right) +} \qquad \textbf{8-6}$$
$$\dot{h} = \alpha_h(1-h) - \beta_h h$$

The Na^{+} conductance is:

$$g_{Na} = 120m^3h \qquad \text{8-7}$$

This system of equations is coded in the program *HHCell*.

HHCell.java

aip.app.hhcell.HHCell *HHCell* is a windowed application with the now-familiar *main()*, *assemble()*, and *run()* methods. Internally, *step()* performs the neuron update calculations as described in Equations 8-1 through 8-7. Figure 8-3 is a screen capture of *HHCell* in action.

The top graph shows the voltage potential across the membrane plus a second trace for the input stimulus. The bottom graph shows the values for the conduction rates *n, m,* and *h*.

The slider controls the external stimulus of the neuron. This stimulus comes in three flavors, a steady-state stimulus, a fixed-frequency pulse with amplitude modulation, and a fixed-amplitude pulse with frequency modulation.

Pulsed Neuron Computing

The HH model is computationally expensive, so while it does a good job of simulating the action potential of a biological neuron, it is also a good

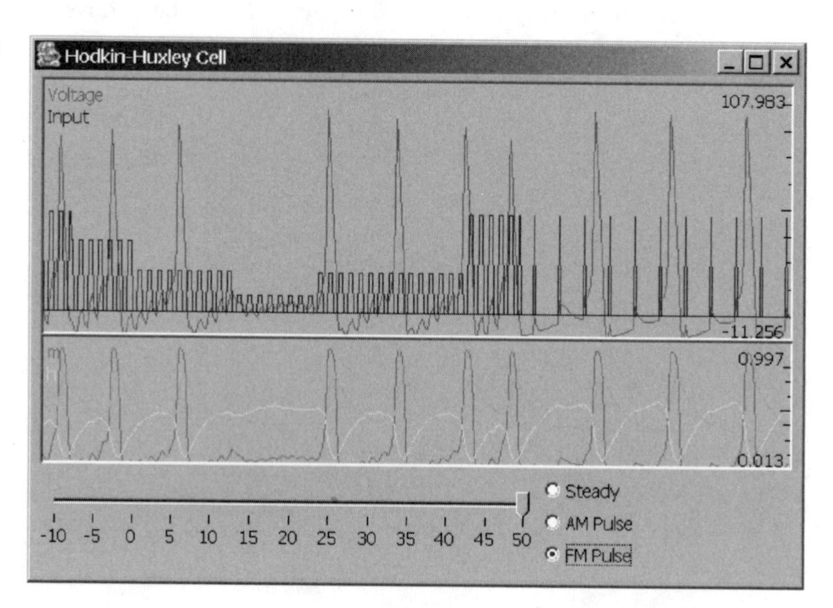

Figure 8-3 HHCell screen capture

way to bring your computer to a grinding halt. Biology has the luxury of operating in true parallel, while our current computer technology tends toward the serial execution of instructions.

Abbott and Kepler (Abbott, 1990) have developed a number of simplified models that preserve the biological dynamics of the HH model. These are still not trivial to compute.

A different approach to modeling neurons is through the use of an oscillator function that can be "pumped" by the dendrite stimulus. An example of this is the FitzHugh–Nagumo relaxation oscillator. Though this model does not accurately mimic all the behavior of the biological neuron, it does model some of it. The math is also considerably simplified, as shown in Equation 8-8.

$$\begin{aligned} \dot{v} &= \varepsilon v(v-\theta)(1-v) - w + I \\ \dot{w} &= (v - \gamma w) \end{aligned} \qquad \textbf{8-8}$$

where

v and w describe a loop in the equation's phase space.

ε is a frequency factor that changes the shape of the phase space of v and w.

θ is a threshold value that controls the sensitivity of the pulse.

I is the input stimulus

γ is a shunt factor, which affects the decay of the pulse.

The FitzHugh–Nagumo oscillator is demonstrated in the program *FNCell.java* in the package *aip.app.fncell.FNCell*.

Of course, you can eschew models entirely and create an arbitrary integrate-and-fire type of system using simple math, such as shown in *AltCell.java* in the package *aip.app.altcell.AltCell*. This, of course, leaves most realism behind in its effort to be easy to compute.

There are many other neural models to choose from, each with its own benefits and drawbacks.

Pulsed neural models provide several advantages over the rate models discussed next, in spite of their increased calculation cost. A number of pulsed-neuron models and computing techniques are explored in Maass (1999), including hardware implementations of spiking neurons.

The values emitted by a pulsed neuron can take two forms. On the one hand, the frequency of the pulses encodes a level of excitement. A voltage integrated across time will be higher with rapid pulses and lower with only a few pulses.

However, you can also use the pulses across a population of related neurons to determine excitement, rather than using just one neuron across time. In this case, the receiving neuron would average the pulses across a short time scale but across all of the transmitting neurons. The more neurons that fire around the same time, the greater the level of excitement from that module.

There are some one-shot models as well. Once a neuron is released from external inhibition, the time delay to the first spike can indicate an excitement level. The sooner it spikes, the more excited the neuron is. Of course, once this spike has been processed downstream, the neuron needs to be reset before any more of its pulses can be considered.

While the excitement level of rate-based neurons is easier to detect than the excitement of pulsed neurons, pulsed neurons provide some unique time-based features. Given an additional timekeeper signal, the pulses from a neuron can code information by their phase relative to the time signal. Groups of neurons can be categorized together if they pulse together.

Receiving neurons can also detect synchronization between two or more inputs, since the pulses exist within a timed matrix.

However, we will not travel any further down this road. The rest of our neural explorations are with rate-coded, or computational, neurons.

Computational Neurons

The type of neuron used most often for computing is the rate-coded neuron, and the most common type of rate-coded neuron uses the McCullock–Pitts model. Rate-coded neurons do not emit pulses but instead send an analog value down their axons and receive these analog values at their dendrites. The neuron itself performs several simple operations in between its input and output, as shown in Figure 8-4.

The inputs to the neuron are the values *x(1)* to *x(n)* which can come from other neurons, external code, or hardware sensors. They are just numbers. Many implementations require that the inputs be in the range *[0..1]* or *[–1..1]*.

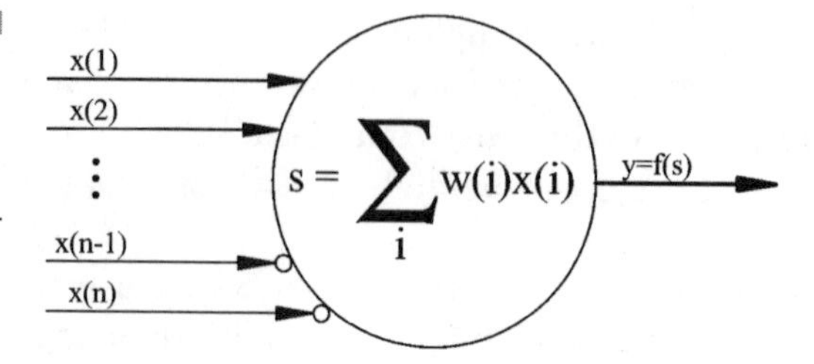

Figure 8-4 McCulloch–Pitts computational neuron

Each input to a neuron has an effectiveness factor, or weight. The input value is multiplied against this weight to determine its influence on the neuron's state. Negative weight values indicate an inhibitory input, and this is often illustrated by drawing a circle at the input's attachment to the neuron. The neuron's state is the sum of the weighted inputs:

$$s = \sum_i w_i x_i \quad \text{8-9}$$

where s is the total stimulation (or inhibition) of the neuron, w_i is the weight, or effectiveness, of this neuron's dendrite input, and x_i is the activation level of the axon output. These weights provide the mechanism for learning; as a network of neurons learn, their weights change; so the neuron's response to input changes. More about this in later sections.

The output of the neuron is then calculated as a function of s. Different models use different output functions.

A Hopfield-type model generates a +1 or –1 output, depending on the sign of s:

$$y = S(s) = \begin{cases} s > \theta : 1 \\ s \le \theta : -1 \end{cases} \quad \text{8-10}$$

where θ is the threshold level above which the neuron triggers, typically 0. Different effects are achieved with positive or negative thresholds. Another standard function is the Heaviside function:

$$y = H(s) = \begin{cases} s > \theta : 1 \\ s \le \theta : 0 \end{cases} \quad \text{8-11}$$

Some applications require a smoothly changing, or differentiable, output rather than a step function. These are the sigmoid functions, one of which is:

$$y = \frac{1}{1 + e^{-s\lambda}} \quad \text{8-12}$$

In this case, the output tends towards 0 as s goes negative, and towards 1 for positive s. The factor λ controls the rate of change, with larger λ making a steeper curve, approaching the step function. λ values less than 1 create a flatter curve. This is shown for λ values of 0.25 – 2.00 in Figure 8-5.

A different sigmoid that goes from –1 to +1 is shown in Figure 8-6, it is based on the tanh*()* function or its equivalent:

$$x = \frac{s}{2}$$
$$y = \tanh(x) \qquad \textbf{8-13}$$
$$y = \frac{e^{x} - e^{-x}}{e^{x} + e^{-x}}$$

The output could even be linear, where $f(s) = s$. There are many different types of activation functions, each with its own particular benefits.

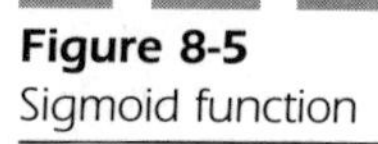

Figure 8-5
Sigmoid function

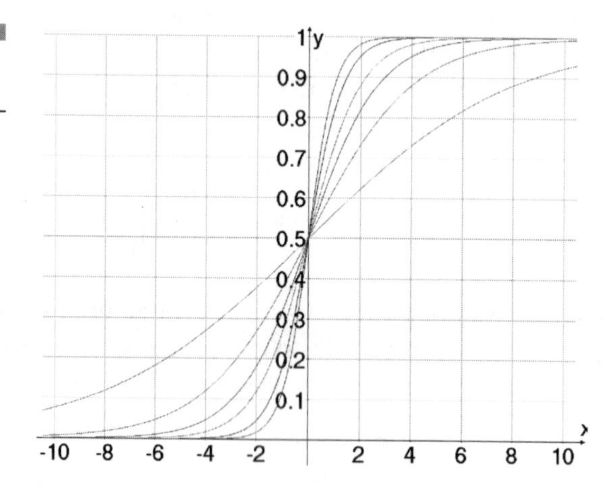

Figure 8-6
tanh() sigmoid function

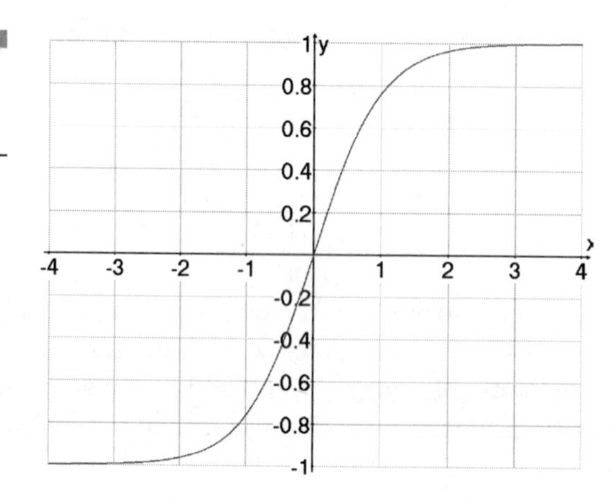

Perceptron

The perceptron is one of the very first uses of neurons as computing elements. It was the brainchild of Frank Rosenblatt, circa about 1957. Using a simple form of the McCullock–Pitts neuron described in the previous section, it learns to categorize input patterns.

The perceptron, and the artificial neural network (ANN) in general, is a form of inductive learning system. They receive input from outside systems and use this data to learn patterns from that data. While the logical systems in the previous chapter would store their knowledge as rules, these networks store their knowledge implicitly in the states and weights of the network's neurons, or neurodes.

The simplest form of the perceptron has one neuron with two inputs, shown in Figure 8-7.

The output of the perceptron is calculated as:

$$y = S\left(\left(\sum_{i=1}^{n} x_i w_i\right) - \theta\right) \tag{8-14}$$

where:

y is the output value
x_i is the value of a given input
w_i is the connection weight for that input
n counts the number of inputs
θ is the bias value, or threshold, of the neuron
$S()$ is the sign function, described in the previous section.

When it is initialized, the connection weights can be set to 0 or some random values. During training, values are applied to the inputs and the result is observed. This result is compared to the expected result and, if it does not match, the weights and threshold, or bias, are updated using Equation 8-15. Presenting both the inputs and the expected answer to train a neuron, or NN, is supervised learning.

$$\begin{aligned} w_i &= w_i + \alpha(T - y)x_i \\ \theta &= \theta + \alpha(T - y) \end{aligned} \tag{8-15}$$

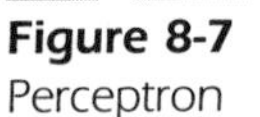

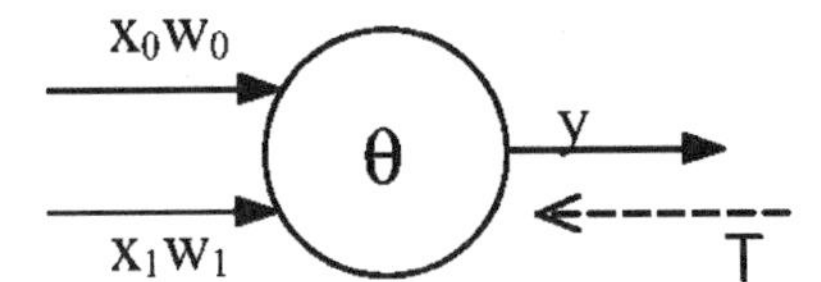

Figure 8-7
Perceptron

where α is a training factor between 0 and 1, and T is the training template, or expected result.

It is convenient to think of the inputs x, the weights w, and later the outputs y, not as a set of numbers but as ordinates in a vector. This, in turn, makes it easier to think of these inputs as describing positions in a state space. In this two-input example, the various allowable values for the input vector can be plotted on a graph, simplifying visualization.

Code: Perceptron

This tests the perceptron, described in Equations 8-14 and 8-15. It is presented by the two classes *Perceptron.java* and *PerceptronXOR.java,* in the package *aip.app.perceptron*.

Perceptron is a command-line program that runs both the two-input perceptron, plus a three-input variation explained under *PerceptronXOR. java* below.

Perceptron.java

aip.app.perceptron.Perceptron.java The implementation of the math in Equations 8-14 and 8-15 is straightforward. Learning the AND function, the perceptron develops the state and weights illustrated in Figure 8-8.

The inputs for the logical AND function are Boolean *true* and *false* which, for this model, are encoded as +1 for *true* and –1 for *false*. Run the calculations by hand and show yourself that the outputs are correct.

You can also change the training data to teach the OR function. However, you may notice that if you train for XOR the program fails.

This leads to the reason that the perceptron was originally abandoned. In 1969, Minsky and Papert published a paper pointing out that the perceptron could only categorize data that was linearly separable. This limitation appeared to be so severe that most work on NNs was abandoned for years.

To be linearly separable, you must be able to separate the data points that are *true* and the data points that are *false* with a single line. It is interesting to note that the two-input perceptron calculation actually

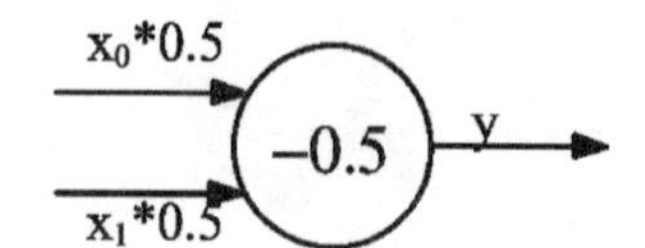

Figure 8-8
AND perceptron

reduces to the equation for a line. All data points that fall on one side of the line are reported as *true*, and the rest are *false*. The math highlights are shown here, starting with the core of Equation 8-14:

$$\left(\sum_i x_i w_i\right) - \theta$$

$$x_0 w_0 + x_1 w_1 - \theta = 0 \qquad \textbf{8-16}$$

$$x_1 = -x_0 \frac{w_0}{w_1} + \theta$$

$$y = mx + b$$

The state space with the decision line for the AND perceptron is shown in Figure 8-9.

The solid circle at *(true, true)* indicates this is a *true* output for those inputs. The dashed line is the line defined by the weights and threshold from Figure 8-8.

Looking at the perceptron this way, it is easy to imagine functions that cannot be divided by a single line. The exclusive-OR operation is the classic example.

PerceptronXOR.java

aip.app.perceptron.PerceptronXOR If you try to learn the XOR function with the standard perceptron, it will fail. However, you can make it work if you increase the dimension of the state space, adding a third input. This input is the AND of the two other inputs, giving the special perceptron shown in Figure 8-10.

This gets around the failure to learn XOR, but it is also cheating. Fortunately, researchers have discovered other techniques that work for all problems, without having to devise special inputs to facilitate learning.

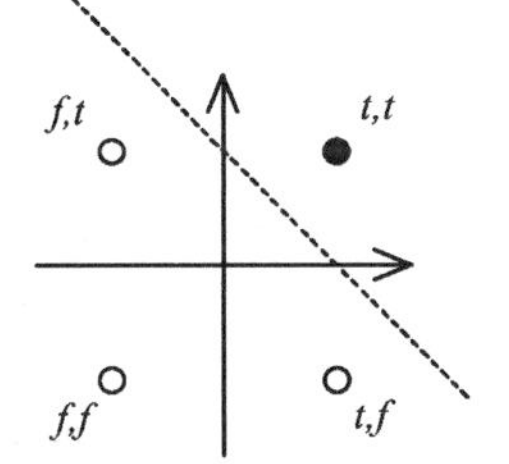

Figure 8-9
AND perceptron state space

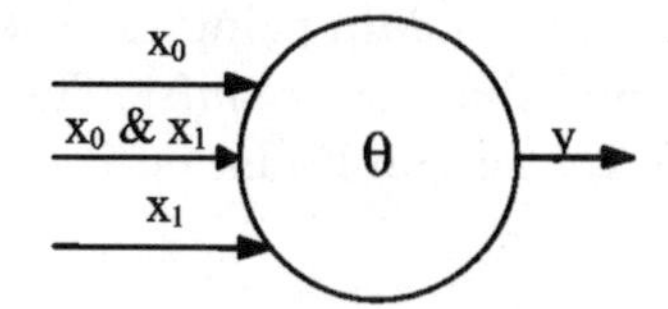

Figure 8-10 XOR perceptron

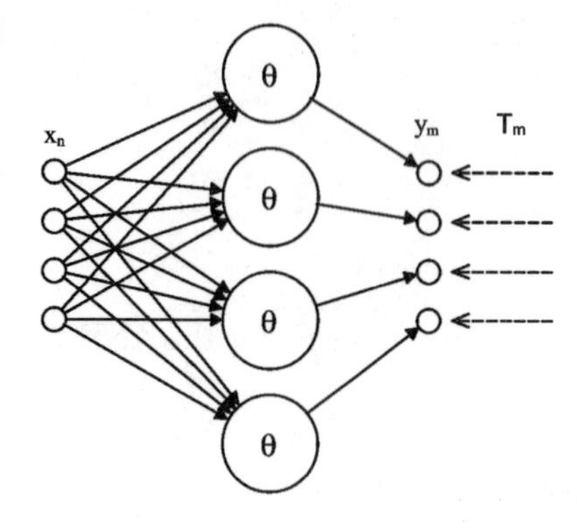

Figure 8-11 Multiple categories

Multiple Categories

The trivial perceptron in Figure 8-7 accurately represents the use of perceptrons in general. In the case where there are more than two categories, the perceptron would be networked as shown in Figure 8-11.

Each neuron can be considered independently, since they do not affect each other. This transforms Figure 8-11 into the set of networks shown in Figure 8-12. The only other difference, then, is the number of inputs. These inputs define the dimension of the perceptron's state space. The categories must still be separable by a flat plane in the input's space.

Multi-Layer Perceptron

The extension that saved the perceptron was the inclusion of an additional layer, creating the Multi-Layer Perceptron (MLP). This architecture, shown in Figure 8-13, is the basis for many NN systems. It is also known as a multi-layer feedforward network, since the information flow travels in just one direction, from the input to the output.

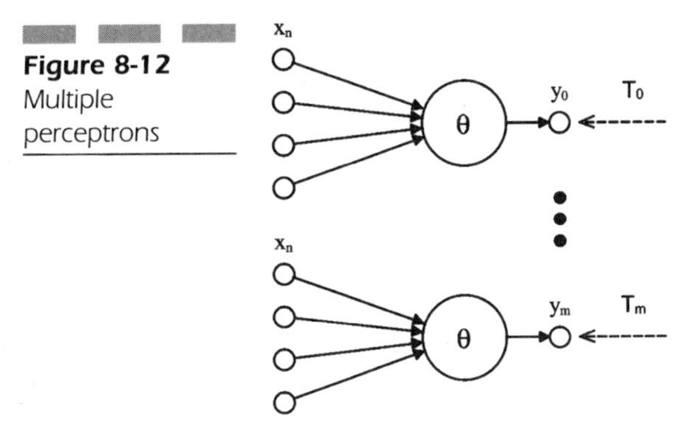

Figure 8-12 Multiple perceptrons

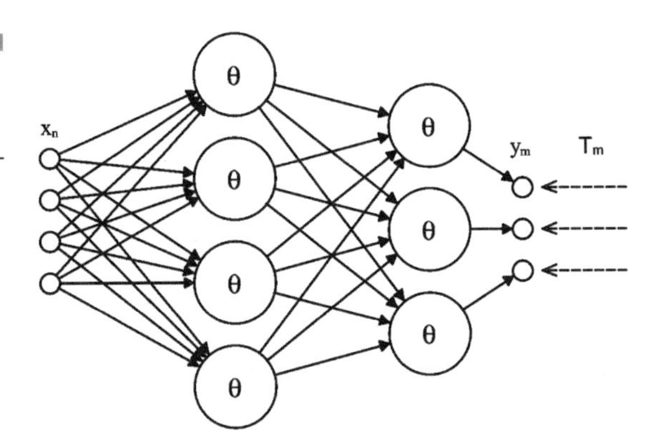

Figure 8-13 Multi-Layer perceptron

This is not to say that there are no other types of neural net to be found. There are many variations on this theme, though if you understand the MLP the other types of network should not be too hard to understand. The MLP covers the basics for this class of network, that is, supervised categorization networks.

The technologies we apply to this network include the delta learning rule and backpropagation of errors, both of which are described later.

Figure 8-13 is considered a three-layer network. The first layer manages the inputs. While the inputs are not neurons perse, they do fully connect to all of the neurons in the next layer. Because of this, it can be convenient to think of them as neurons with input weights of 1.0 and thresholds of 0.0.

The next layer is the "hidden" layer. It is hidden because it does not connect directly to the output, or get feedback directly from the training vector. It is hidden behind the third layer.

The third layer contains the output neurons.

Networks may have more than three layers, but you rarely need these additional layers. More layers slow down learning, and even three-layer networks tend to have learning cycles of hundreds or even thousands of presentations.

Both the hidden and the output layers need to be trained before the network can do work.

Many NNs use a three-layer architecture. The hidden layer is powerful enough to allow the network to learn almost any type of function, given the right number of neurons. However, the step output function used in our previous perceptron does not work very well for training the middle layer. We need to use a different output function, such as one of the sigmoid functions.

Choosing the right number of neurons in the hidden layer is important. Too many and each neuron "memorizes" one condition, losing the ability to generalize. Too few and it will not learn the function. A rule of thumb says that if there are n input neurons and m output neurons, the hidden layer should begin with $\sqrt{mn}$ neurons (Masters, 1993).

State Space

The perceptron example was in a simplified state space. The inputs and the outputs were locked at the extremes of plus or minus one. Most networks operate in a subtler environment.

We talked about the input values as describing a vector. This is even more true when the inputs have arbitrary real values.

The set of weights for a neuron can also be considered a vector. The multiplication of the input by the weights is then the same operation as the dot, or inner, product. This operation, if you recall your math, returns the cosine of the angle between the two vectors times the length of those vectors. The dot product is a measure of how close the vectors are to being the same.

Learning, then, is the task of bringing the vectors together. Note that not all neurons learn all vectors; the task is spread around. But if a neuron's weight vector is already close to the input vector, it is moved to be even closer.

Delta Rule

In our earlier perceptron the learning rule was a simple linear shift towards the desired result. Linear learning rules, however, do not work in the multi-layer network. In fact, multiple layers lose their advantages if you are using linear learning rules (Kremer, 1999).

The delta rule improves on the linear learning rule, making the weight motion follow the curve of a hyperboloid, weight changes are larger when they are far from the target and smaller as they get closer.

The delta rule is also known as the least-mean squared error rule, LMS, or the Widrow–Hoff rule, after its discoverers. The rule itself is

$$\bar{w} = \bar{w} + \frac{\alpha(T - y)\bar{x}}{|\bar{x}|^2} \qquad \textbf{8-17}$$

While this looks eerily the same as the weight update rule from Equation 8-15, it has some significant differences.

$\bar{w}$ is the weight vector for this neuron
$\bar{x}$ is the input vector
$|\bar{x}|^2$ is the length of the input vector squared.

The other values, T, y, and α are all the same scalars that we know and love.

The delta rule applied to the perceptron provides some additional power to that system, however, it is most often used in conjunction with sigmoid activation functions in multi-layer networks using backpropagation of errors.

Backpropagation

Using a nonlinear update rule is only the first step in making a MLP. If you look at Figure 8-13, you can see how the output layer can be trained. The activation signals from the hidden layer are processed, and the generated results of the output layer are then correlated with the training pattern.

But how does the middle, hidden, layer get trained?

The trick is to propagate the error signal from the output layer back to the middle layer, hence the name backpropagation. When a neuron is generating an incorrect result in the output layer, it may be because of its input weights. But the data coming from the middle layer may also be wrong, so we need to pass some of this blame upstream to the middle neurons.

But what *is* the error term at the middle layer? For the output layer, the error is the difference of the actual output from the desired output. But one step removed from the actual output and training signal, it is harder to know what the error is.

Let us adjust our annotation a bit to compensate for the complexity of the three-layer network. The input neurons have fixed values x_i that feed through weights w_{ij} to the middle layer neurons. These neurons calculate activation levels s_j from this input. The neurons have the internal threshold values θ_j, error values e_j, and outputs y_j, which become the output layer's inputs, x_k. These outputs pass through weights w_{jk} to the output neurons, which have threshold values θ_k, activation levels s_k, error values e_k, and outputs y_k. The training value is still T. The layers, essentially, are identified by their subscripts i, j, and k.

The error sent back to the middle layer is moderated by the impact this layer's output had on the output layer. To do this, e_k is multiplied by the weight factor between the middle and output neurons w_{jk}:

$$e_j = \sum_k w_{jk} e_k \qquad \textbf{8-18}$$

where

e_j is the error for a neuron in the middle layer.

w_{jk} is the connection weight between neuron j in the middle layer and neuron k in the output layer.

e_k is the error signal at neuron k, which is $(T_k - y_k)$ from Equations 8-15 and 8-17.

Recalling the discussion on state space, it makes sense to adjust a neuron that is mostly undecided rather than neurons that are already very close to the input vector or are very far from it. So neurons with very high or very low outputs do not shift very much, while "undecided" neurons get adjusted more. An easy way to control this is to use the derivative of the activation function as a factor of the error signal. Of course, to get a derivative we must have a sigmoid or other smooth activation function:

$$\begin{aligned} y_j &= \frac{1}{1+e^{-s_j\lambda}} \\ y_j' &= y_j\lambda(1-y_j\lambda) \\ e_j &= y_j'\left(\sum_k w_{jk}e_k\right) \end{aligned} \qquad \textbf{8-19}$$

The sigmoid function y with its derivative y' is shown in Figure 8-14.

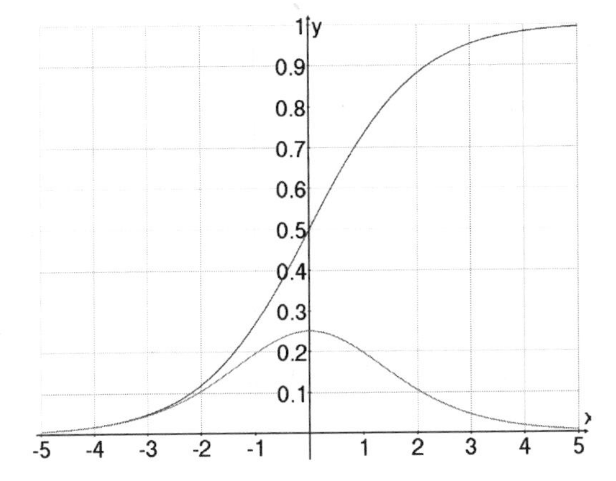

Figure 8-14 Sigmoid and derivative

Summary So to recap, an input pattern x is presented at the input layer of the MLP. The activation level of the middle neurons are calculated using a modified form of Equation 8-9, which takes the threshold into account:

$$s_j = \left(\sum_i w_{ij} x_i \right) - \theta \qquad \textbf{8-20}$$

The output of each neuron in the middle layer is then calculated using the sigmoid from Equation 8-12, which is

$$y_j = \frac{1}{1 + e^{-s_j \lambda}}$$

These outputs are sent as inputs to the next layer, and those neurons perform the same calculations, generating outputs y_k.

The ultimate output y_k is compared to the training signal T_k to get an error term e_k at the output layer.

The output layer is then trained:

$$\begin{aligned} e_k &= T_k - y_k \\ w_{jk} &= w_{jk} + \alpha e_k y_j \\ \theta_k &= \theta_k - \alpha e_k \end{aligned} \qquad \textbf{8-21}$$

The error term calculated at the output neuron is then fed back up to the middle layer neurons:

$$y'_j = y_j\lambda(1 - y_j\lambda)$$
$$e_j = y'_j\left(\sum_k w_{jk}e_k\right) \qquad \textbf{8-22}$$

and then the weights in the middle layer are trained:

$$\dot{e} = \frac{\alpha e_j}{|\bar{x}|^2}$$
$$w_{ij} = w_{ij} + \dot{e}y_j \qquad \textbf{8-23}$$
$$\theta_j = \theta_j - \dot{e}$$

Training Even after the network has been built, special care needs to be taken in training it.

Multi-layer backpropagation networks train slowly, so you need to present the input/output training data many times for it to take hold. The perceptron example earlier in this section is not representative of the difficulty of training NNs! How many times you need to present the data depends on your specific problem. The overall level of the error values indicates how close the network is to perfection. The network will rarely, if ever, reach zero error, but it will settle into some low value.

Note that while NNs can be slow to train, they calculate their results quickly once trained.

The training examples need to be representative of the real data the network will process. While neural nets can adapt to noise, and can do some interpolation between the trained exemplars, they do not do a very good job of managing inputs that are entirely outside of the trained state space.

Also, the order in which you present the training data is important. If you give the first example several hundred times, then the second example, and so forth, by the time you reach the last examples, all earlier knowledge will have been forgotten. It is important to mix up the training data, presenting them randomly to minimize this forgetting.

Once the network is trained, it should be tested against data that was not in the training set, to see how it performs. If it performs poorly you can continue to train the network.

New examples can be trained into the network during the network's operation, though it is beneficial to retain the old training examples and present them with the new ones, to minimize forgetting.

Code: Backpropagation Character Recognition

Let us take our knowledge of backpropagation networks and create a simple character recognition program.

To avoid the complexities of a user interface to capture character data, this program operates with test data generated outside of the application and stored in GIF files. These characters were drawn on paper, grabbed in a graphics program, and scaled to fit a 12 × 16 grid. Of course, in practice the characters will be grabbed, framed, scaled, and otherwise adjusted automatically by some other input software.

This program cheats to an extent, as well. It tests against the data that was used to train the network. This test program performs its recognition task fairly well, even in the face of a noisy input, but it does not generalize particularly well. If you test with data that was not in the training set, for these sloppy letters, the quality of the recognition goes down. There are different ways this might be solved; different parameters for the network, more training examples, *better* characters, different output format, and so on. The challenge of NNs is in finding just the right combination of factors and training environment to get it to perform the way you want it to. It is as much an art as it is a science.

One problem to watch for during training is overfitting. If you have too few examples, too many middle-layer neurons, or run too many training loops with too high a learning rate, you run the risk of making the network too specific to operate on data other than the examples presented to it. Instead of being a nice, loose-fitting categorizer that can gracefully manage a wide variety of input data, it snaps into a close fit around the test data and makes a mess of, or rejects, other input data.

The class diagram for this test program is given in Figure 8-15.

Vector.java

aip.math.Vector The *Vector* math class stores an array of double values for its ordinates, and performs some basic math on them. You can add, subtract, and scale vectors, as well as get their length and the dot product of two vectors.

A summary of *Vector* is given in Table 8-1.

Perceptron.java

aip.neuron.Perceptron The *Perceptron* class is an abstract base class that gives you the framework to represent a single layer of neurons.

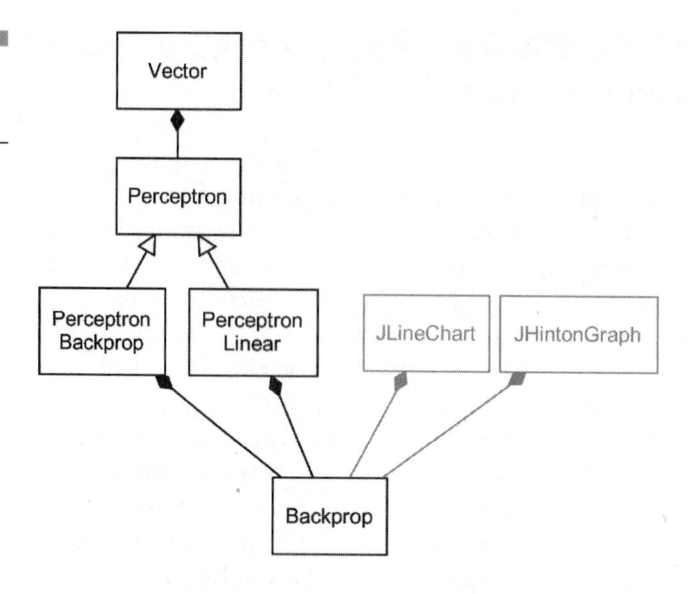

Figure 8-15 Backprop class diagram

Table 8-1 Vector summary

Construction	
	Vector (int dim)
	Vector (Vector vec)
	Vector (double[] ord)
void	init (Vector vec)
void	init (double[] ord)
Operations	
void	add (Vector vec)
void	sub (Vector vec)
double	length ()
void	scale (double factor)
void	scale (Vector vec)
void	normalize ()
double	dot (Vector vec)
Access	
int	getSize ()
double	getOrd (int idx)
void	setOrd (int idx, double ord)

The structural aspects are managed here, but the learning and computing methods are left undefined.

A summary of *Perceptron* is given in Table 8-2.

Perceptron(int num) The constructor allocates all of the storage vectors except the weight vector, and it sets the thresholds to zero. Each vector is the same length, as specified in the *num* parameter.

Table 8-2

Perceptron summary

```
Construction
                        Perceptron (int num)
                  void  initWeights (Vector invec)

Operations
abstract Vector  calculate (Vector input)
  abstract void  train (Vector template, double alpha)
  abstract void  train (Perceptron template, double alpha)

Access
         Vector  getInput ()
         Vector  getState ()
         Vector  getThreshold ()
         Vector  getError ()
         Vector  getOutput ()
       Vector[]  getWeight ()
         Vector  getWeight (int node)
```

initWeights(Vector invec) Using connection weights is optional in the bigger picture, so the weight initialization is broken out into its own method. The specified *invec* is used to determine the size of the weight vectors. Once the weight vectors have been allocated, they are initialized to a random value in the range [–1.0 ... +1.0].

When using input weights, the input vector may be a different dimension than the perceptron layer, so the input vector is re-defined to be the same size as the *invec*.

getInput()
getState()
getThreshold()
getError()
getOutput()
getWeight()
getWeight(int node) The access methods return the many internal vectors that make up the perceptron layer.

PerceptronLinear.java

aip.neuron.PerceptronLinear This is a specific implementation of the *Perceptron* base class that uses a linear learning law and a linear output function. The Heaviside or sign function can be applied outside of the *PerceptronLinear* when processing the output, if desired.

A summary of *PerceptronLinear* is given in Table 8-3.

Table 8-3

PerceptronLinear summary

```
Construction
              PerceptronLinear (int num)

Operations
       Vector calculate (Vector input)
         void train (Vector template, double alpha)
         void train (Perceptron template, double alpha)
```

calculate(Vector input) Processes the input vector, generating an output vector. If there are no weights, the input is copied directly to the state:

```
m_state.init(input);
```

If there are weights, the input is factored against each node's weights to get a state value. This is an implementation of Equation 8-9:

```
for (int idx=0; idx<num; idx++)
{ m_state.setOrd(idx, input.dot(m_weight[idx])); }
```

train(Vector template, double alpha) This method trains the weights of the perceptron, using the training rate *alpha*, to make it more likely to generate the *template* vector. This is an implementation of Equation 8-15:

```
m_error.init(template);
m_error.sub(m_output);

adjustWeight(alpha);
```

The private *adjustWeight()* method continues the calculation:

```
Vector f_error = new Vector(m_error);
f_error.scale(alpha);   // alpha*e
//
//
int num = m_state.getSize();
for (int idx=0; idx<num; idx++)
{
  double error = f_error.getOrd(idx);
  //
  // w = w + x*(alpha*e)
  //
  Vector e_input = new Vector(m_input);
  e_input.scale(error);
  m_weight[idx].add(e_input);
}
// t = t - (alpha*e)
//
m_thresh.sub(f_error);
```

train(Perceptron template, double alpha) This form of training is not defined for *PerceptronLinear*, so it simply throws an *IllegalAccess Exception*.

PerceptronBackprop.java

aip.neuron.PerceptronLinear This is a specific implementation of the *Perceptron* base class that uses the delta learning law and a sigmoid output function.

A summary of *PerceptronBackprop* is given in Table 8-4.

calculate(Vector input) Processes the input to set the perceptron's internal states and generates an output vector. If no weights have been defined, it simply copies the input to the state, otherwise the calculation implements Equations 8-20:

```
m_input.init(input);
for (int idx=0; idx<num; idx++)
{ m_state.setOrd(idx, input.dot(m_weight[idx] - m_thresh.getOrd(idx)); }
```

The output is calculated according to Equation 8-12:

```
for (int idx=0; idx<num; idx++)
{ m_output.setOrd(idx, 1.0 / (1.0 + Math.exp(-m_state.getOrd(idx)))); }
```

train(Vector template, double alpha) This method trains the perceptron layer against the specified *template* vector. The error is calculated directly:

```
m_error.init(template);
m_error.sub(m_output);
adjustWeight(alpha);
```

The weights are adjusted as described later.

train(Perceptron template, double alpha) The *template* perceptron is the downstream output perceptron layer. This training method uses the error vector from this layer to determine its own weighted error level according to Equation 8-22:

```
Vector error = template.getError();
int knum = error.getSize();
```

Table 8-4 Perceptron-backprop summary

Construction

```
                PerceptronBackprop (int num)
```

Operations

```
         Vector calculate (Vector input)
           void train (Vector template, double alpha)
           void train (Perceptron template, double alpha)
```

```
int jnum = m_state.getSize();
for (int jdx=0; jdx<jnum; jdx++)
{
   // e = y'(sum of weighted errors)
   //
   double sum = 0.0;
   for (int kdx=0; kdx<knum; kdx++)
   {
        sum += error.getOrd(kdx) * m_weight[kdx].getOrd(jdx);
   }
   double out = m_output.getOrd(jdx);
   double deriv = out*(1-out);
   m_error.setOrd(jdx, sum*deriv);
}

adjustWeight(alpha);
```

Then the weights are adjusted in the private *adjustWeight()* method as per Equation 8-23:

```
double factor = alpha / (len*len); // alpha / |e|^2
Vector f_error = new Vector(m_error);
f_error.scale(factor);  // alpha*e / |e|^2
//
//
int num = m_state.getSize();
for (int idx=0; idx<num; idx++)
{
   double error = f_error.getOrd(idx);
   //
   // w = w + x*(alpha*e / |e|^2)
   //
   Vector e_input = new Vector(m_input);
   e_input.scale(error);
   m_weight[idx].add(e_input);
}
// t = t - (alpha*e / |e|^2)
//
m_thresh.sub(f_error);
```

Backprop.java

aip.app.backprop.Backprop The *Backprop* test program provides an intricate user interface to support the process of training, and then testing, the MLP model.

Most of the work is in creating and managing this extensive UI, through a nice bit of code loads and converts the GIF files into various forms. The interested reader can scan the source code for the details. The heart of the application lies in the two methods *do_train()* and *test()*.

Training is a simple matter of running an input through the network and then training each layer:

```
m_input.calculate(invec);
m_middle.calculate(m_input.getOutput());
m_output.calculate(m_middle.getOutput());

int code = 'A' + letter;
loadTemplate(outvec, code);

m_output.train(outvec, m_rate);
m_middle.train(m_output, m_rate);
```

The *test()* method is even simpler, simply running an input through the network and collecting the results.

The user interface for *Backprop* is shown in Figure 8-16.

Training Area The top area shows the test data, in this case four sets of capital letters. And yes, my printing is this bad. Below the test data is the error graph which keeps you up to date on how well the network is performing.

If the training data does not show up, the *Backprop* data loader is probably not finding the GIF files. Try adjusting the data path specified at the top of *Backprop.java*:

```
final private static String data_path = new
String("aip/app/backprop/data/");
```

Figure 8-16 Backprop user interface

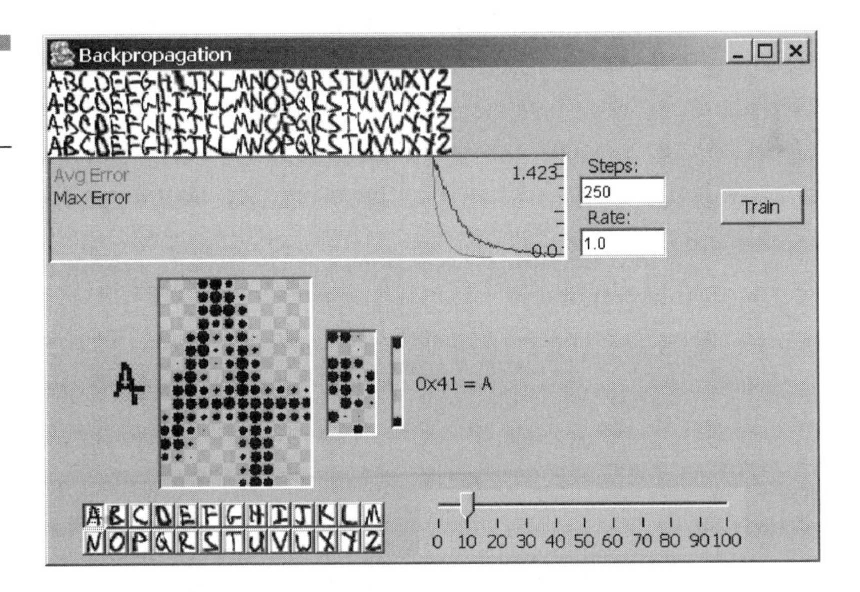

To the right of the error graph are the training controls. *Steps* determines how many training steps to perform each time the *Train* button is pressed. One step is a single pass through the entire set of training data. In addition, each training pass randomizes the order in which the training data is presented to the network.

The *Rate* input determines the learning rate of the network. It defaults to one, since the delta rule has a built-in learning rate factor; the closer the output is to the preferred result, the slower the network trains. Of course, you can experiment with different training factors.

Network Status The network status is displayed next. The first image is an enlarged view of the test character, displayed against a threshold of one half. The next image is the output from the input layer. This shows the input data in its more subtle form, with shades of gray represented by different sized dots in the Hinton diagram. Any negative inputs, because of noise, would be red.

The third Hinton diagram shows the output from the hidden layer. This is essentially gibberish.

The fourth and final diagram is the output layer. This is the lower seven bits of the character's ASCII code, with bit zero at the bottom. To the right of this, the code is interpreted against a threshold of one half.

The use of ASCII code as our output is unusual. A more common form for categorization problems is to dedicate a single output bit to each possible category, bit zero for "A", two for "B", and so on.

The way the network displays its test data is deceptive... it looks like characters! As such, when we watch the network process the input data, we use our own intuition as to whether it should recognize an input or not. As likely as not, our intuition will be wrong.

The network has no understanding of characters, lines, shape, form, or anything. The input data is just a string of numbers that it is performing certain numerical operations on. The results are equally meaningless to the computer.

To get a better sense of what the computer has to work with, turn on *OBFUSCATE* mode and compile and run the program again. Figure 8-17 is what you will see for the letter "A", for example.

```
final private static boolean OBFUSCATE = true;
```

Meaningless hash! But this is what the input means to the computer; nothing. With no understanding of the underlying form, it is no wonder

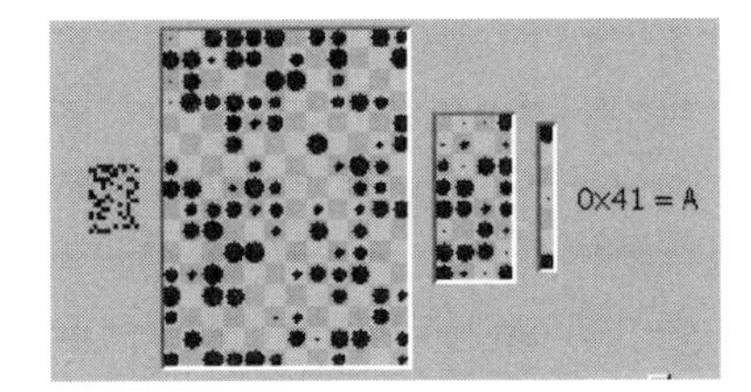

Figure 8-17
Obfuscation

the computer fails to recognize the characters sometimes, at least when enough noise has been added.

Test Area At the bottom of the window are two rows of buttons, each with a letter on it. Pressing a button randomly selects one of the versions of that letter and sends it through the network. Note that the buttons can be pressed at any time, even during training.

The slider to the right of the buttons controls the noise levels in the input pattern. Note that the network rejects noise better if it is trained with some noise. The noise slider adds noise during both training and testing.

Momentum

Imagine, for a moment, what the error landscape of the MLP looks like. The error during the first iteration is fairly large, but then the network trains. This brings the network closer to the goal, so the next time that pattern is executed the error is smaller. Over time the weights move until the error is at the lowest point possible. Any weight changes from this point increase the error, so the weights eventually settle back down to the low point, like a marble settling in the bottom of a bowl (Figure 8-18).

Unlike a marble, however, the movement of the weight vector has no mass, no momentum. If the fitness landscape is not a smooth curve, but has lumps and wiggles in it, like Figure 8-19, the network will stop learning once it reaches a local minimum. If the weight vector did have momentum, like a solid marble, then perhaps that momentum could carry it forward, up and out of the depression and on to a global minimum as shown in Figure 8-20.

To apply momentum to the learning process, you need to record another value with each weight. Each weight needs to remember how

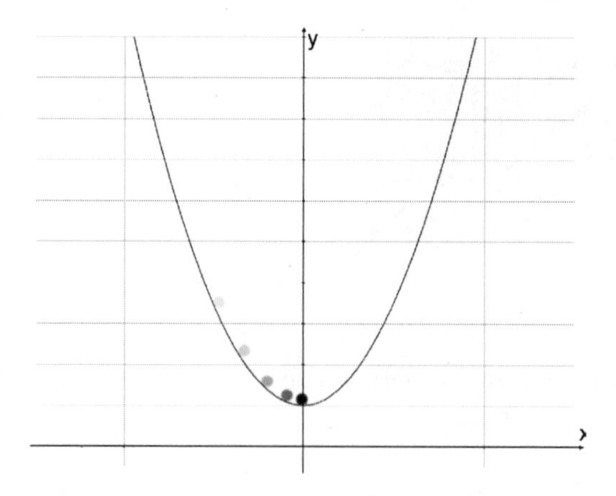

Figure 8-18
Fitness landscape

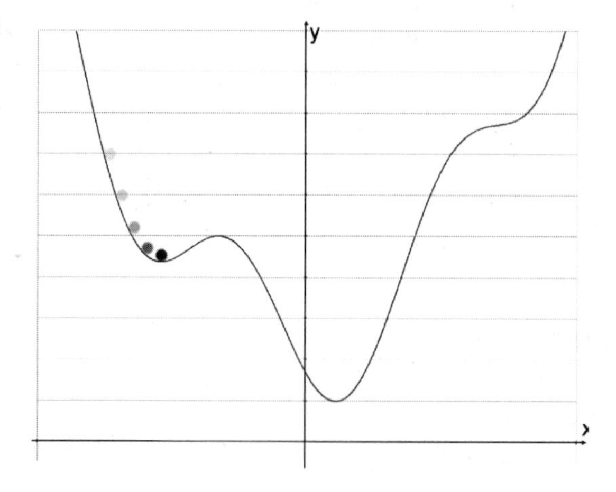

Figure 8-19
Local minima

much it changed during the previous training pass. Revisiting Equation 8-21, we can incorporate the momentum information:

$$\delta = \alpha e_k y_j$$
$$w_{jk} = w_{jk} + m_{jk} + \delta \quad \textbf{8-23}$$
$$m_{jk} = \beta\delta$$

where

δ is a placeholder for the weight change

m_{jk} is the momentum for this weight

β is the momentum factor. This may be any value from 0 to 1, or it could be based on α, e.g. $\beta = (1.0 - \alpha)$.

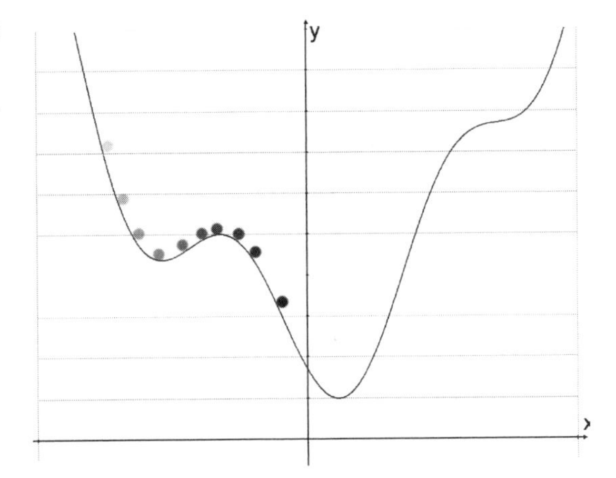

Figure 8-20
Momentum

After the momentum from the last weight update has been used, it is recalculated based on the current weight motion.

The momentum values are provided for in the *Perceptron* class, but are not used in this implementation of *PerceptronBackprop*.

Normalization

The inputs and outputs of your NN may not be useable as-is, but may need to be adjusted, or normalized. Of course, the subject of data normalization is not limited to NN use. Many algorithms benefit from massaging the input or output data to match a standardized format.

Even in the simplified character recognition program in the previous section, the input characters were not used raw. After they were scanned off the paper, I adjusted their color saturation and contrast, captured their bit pattern, and scaled them to fit the 12×16 input area.

The normalization process is used to remove irrelevant information from the data, as well as to enhance or extract its significant features. In some cases, the data can be reduced in size or dimension to make the downstream processing faster. The specific processes that can be applied to the input or output data depend on the problem being solved.

There are many different ways you can manipulate data, including statistical techniques like principle component analysis and others. We look at several fairly simple techniques here.

Filtering

A time-varying input signal can be run through a filter to reduce the low or high-frequency components. A simple filter averages several inputs into one. Other filters have been explained in Chapter 3.

Scaling

Data values can fall all over the map. You usually *want* them to fall between well-behaved limits, typically in the range [0 . . . 1] or [−1 . . . +1]. Though the data could be a group of unrelated values, it is easier to think about them as the different ordinates of an input vector.

Ordinate Scaling Simply scaling the input vector is one way to approach the problem. Normalizing the ordinates of a vector into an arbitrary range is a matter of finding the largest and smallest ordinates and using them to adjust the vector:

$$x_i' = \left(\frac{x_i - \min(x)}{\max(x) - \min(x)} \right) (\text{out}_{\max} - \text{out}_{\min}) + \text{out}_{\min} \qquad \textbf{8-25}$$

where

x_i is the original value, and x'_i is the scaled value.
$\min(x)$ is the value of the smallest ordinate in the original vector.
$\max(x)$ is the value of the largest ordinate.
$\text{out}_{\min}$ and $\text{out}_{\max}$ are the bounds of the destination range.

Vector Normalization Instead of making the individual ordinates fall into a specific range of values, you may need to make the vector itself have unit length. By normalizing all of the vectors in a system, you constrain it to a state space that is a circle with a radius of one, and all dot product operations give the true angular difference between vectors.

Normalizing a vector consists of dividing each ordinate by the length of the vector itself:

$$x_i' = \frac{x_i}{|x|} \qquad \textbf{8-26}$$

where the length of the vector is calculated with our trusty Pythagorean theorem:

$$|x| = \sqrt{\sum_i x_i^2} \quad \textbf{8-27}$$

Statistical Z-Score Scaling The ordinates can be scaled so that they are ordered relative to their distribution, using the statistical mean and standard deviations explored in Chapter 1:

$$\bar{X} = \frac{\sum_i X_i}{N}$$
$$s = \sqrt{\frac{\sum_i (X_i - \bar{X})^2}{N-1}} \quad \textbf{8-28}$$
$$x_i' = \frac{x_i - \bar{X}}{s}$$

Of course, you need to pick reasonable samples for *X* for this to work.

Sigmoid The sigmoid functions, used earlier in this chapter for scaling the activation value into a well-behaved output value, can be applied to data in other places as well. To keep this topic from being too pitiful, here is a sigmoid function that scales to the range [–1 .. +1]:

$$x_i' = \frac{1 - e^{x_i}}{1 + e^{x_i}} \quad \textbf{8-29}$$

The input to the sigmoid could be the raw data or, for that extra tangy flavor, the statistically normalized *z*-score value. Running the *z*-score value through the sigmoid compresses any wild numbers at the extreme edges of the bell curve into an acceptable range.

Z-Axis Normalization

One problem with many of the scaling methods above is that the magnitude information in the vector is lost. The vector *(0.001, 0.000)* is the same, after scaling, as *(1.0, 0.0)*. This might radically change its meaning. It depends, as usual, on your problem space.

If you want to preserve magnitude, but still want normalized vector lengths, *Z*-axis normalization can help. Note that the *Z* in this scaling technique has nothing to do with the *Z* from the statistical processes. It refers, instead, to the *Z* axis of three-dimensional space, assuming your input vector is using the *X* and *Y*-axes. Of course, this technique also applies to input dimensions greater than two.

Given an input vector with n ordinates, we add a new synthetic ordinate at $n+1$. All of the ordinates are then scaled so that the vector has a length of one. Note that, before we begin, the ordinates need to be in the range of 0 to 1, otherwise the *Z*-scaling doesn't work. Sigmoid pre-scaling would shift the data into this range, preserving their relative magnitudes.

The scaling factor is

$$f = \frac{1}{\sqrt{n}} \qquad \textbf{8-30}$$

where n is the number of ordinates in the raw vector. This is then applied to the ordinates of the data vector:

$$x_i' = fx_i \qquad \textbf{8-31}$$

The new ordinate is then calculated to make the entire vector return a length of one:

$$x_{n+1}' = f\sqrt{n - |x|^2} \qquad \textbf{8-32}$$

where $|x|^2$ is the length of the original data vector.

Binary Representations

You can often apply a real-valued signal directly to an input on a neuron. This is not always desirable or even possible. For example, some neural net models are binary and do not understand the many shades of grey between 0 and 1. In these cases, there are two basic ways to convert a real value to binary inputs for the network.

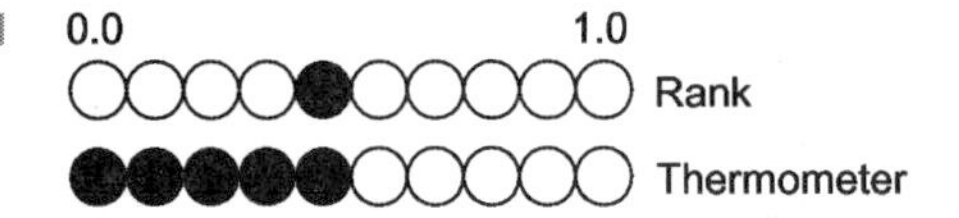

Figure 8-21 Rank interval codings

Neither of these techniques use the binary code of the number directly.

One real-to-binary conversion provides a number of bins or rank intervals, one interval per bit. Whichever bin the input value falls into gets a *true* value and the rest get *false*.

Another conversion is essentially the same as the rank interval conversion, except that all of the bins at and below the input value are set to *true*. This is the thermometer representation. These codings are illustrated in Figure 8-21.

In both these schemes, the intervals could be spread evenly between the min and max values of the input, or the spacing could be logarithmic. Note that nature tends to prefer logarithmic solutions.

Softmax

When a rank interval or category flag is used as the output of a network, one bit is significant and the others should fade into the background. One way of selecting the significant bit is with the *max()* function, selecting one bit to be *true* and the rest to be *false*. However, your output may simply be an intermediate step in a series of operations and the *max()* function may not fit into the scheme of things.

Softmax scaling normalizes the output at the same time as it enhances the differences between its ordinates. This is similar, but different, from the softmax described for reinforcement learning in Chapter 6. This softmax is related to the fitness scaling function from Equation 6-21:

$$y_i' = \frac{e^{y_i/T}}{\sum_n e^{y_n/T}} \qquad \textbf{8-33}$$

where T controls the exponential curve. T values of less than 1 enhance the differences. The sum of all ordinates will be one.

Associative Memory

The networks explored so far in this chapter are, technically, associative memories. They associate an input pattern with an output pattern. However, feedforward networks are traditionally thought of as function mapping networks instead, taking an input vector and mapping it to an output vector. The distinction may seem irrelevant, however, associative memory networks use a different structure and learning method than the feedforward networks.

Bidirectional Associative Memory

Bart Kosko's bidirectional associative memory (BAM) is similar to the multilevel perceptron, with some differences. It is a two-layer network as show in Figure 8-22.

The weights between the layers are shared and symmetrical, so that $w_{ij} = w_{ji}$. The neurons themselves generate output values of +1 for *on* and −1 for *off,* based on inputs to that neuron:

$$y_i = S\left(\sum_j w_{ij} x_j\right) \qquad \textbf{8-34}$$

where

y_i is the output of a neurode.

x_j is an input to a neurode, which is an output from the other layer.

w_{ij} is the weight of the connect between the two neurodes.

During operation, a pattern is presented at one side, A or B, and the signal propagates to the other side. The pattern that is generated is sent back through the weights to the original side. This oscillation, or resonance, is repeated until the pattern stabilizes. Sometimes, however, the network may oscillate between two patterns instead of settling down into one.

The Hebbian learning law is used to teach the network. The weight values are set directly, with each pattern pair directly modifying the weights:

$$w_{ij} = w_{ij} + \frac{1}{N}\sum_{n=1}^{N} x_i[n] x_j[n] \qquad \textbf{8-35}$$

where

w_{ij} is the weight between two neurodes. It is initially 0.

N is the number of neurodes j that connect to this neurode i.

x_i and x_j are the pattern presented to the neurodes on each layer.

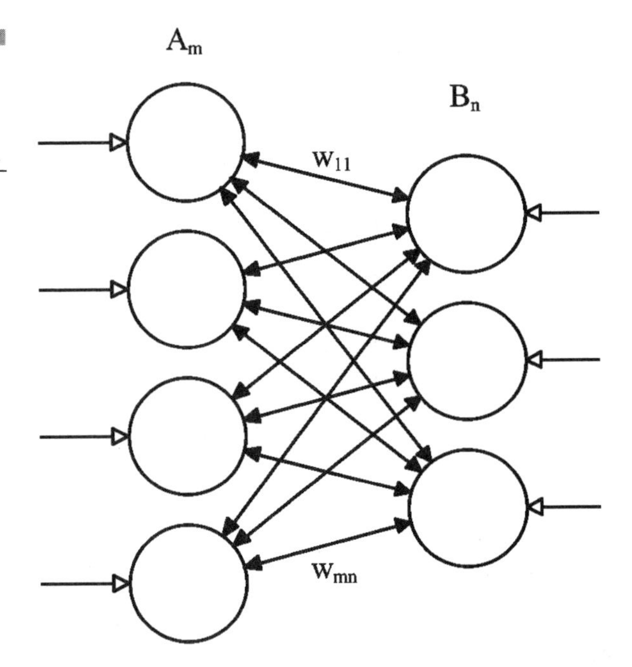

Figure 8-22 Bidirectional associative memory

Hebbian learning is one of the first learning rules, developed by Donald Hebb based on his observations of nature. Simply put, if the output of one neuron matches the state of its destination neuron, the connection weight between the two is strengthened. You will note that if x_i and x_j are both +1 or both −1 then the weight is strengthened, otherwise it is weakened.

Code: Hopfield Network

A Hopfield network is like a BAM network where the input and output layers are of the same size. With layers the same size, and with symmetric weights, it is as if the two layers were really only one layer. So the Hopfield network can, in fact, be reduced to a single layer of neurodes, each connected to every other (Figure 8-23).

The weight from a neuron to itself is 0, $w_{ii} = 0$, otherwise the Hopfield net operations and learns the same way as the BAM.

A common use for Hopfield networks is in the reconstruction of partial patterns. Once a pattern has been trained into the weights, you can present a noisy piece of this pattern and the rest of it can be regenerated.

Hopfield networks need at least seven neurodes for every pattern it is asked to memorized, though for certain well-behaved cases it can do with as few as four neurodes per pattern. Also, the patterns that work the best

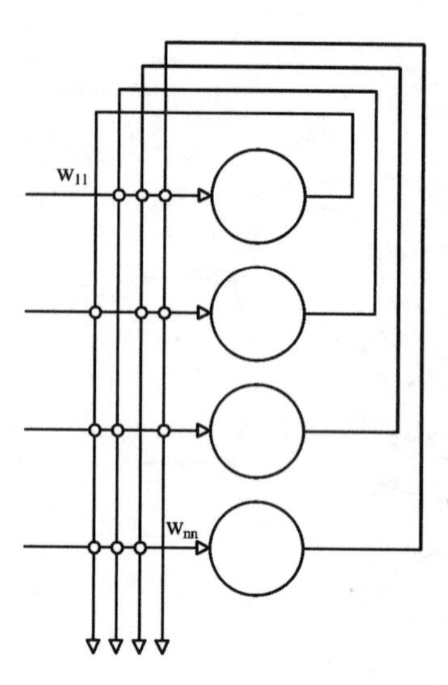

Figure 8-23
Hopfield network

have roughly the same number of *on* versus *off* neurodes. A minimum of overlap between the patterns also helps keep the network from getting overly confused.

Training a pattern into a Hopfield network is a one-step action. The weights are calculated according to Equation 8-35. Execution is the familiar Equation 8-34.

NetworkHopfield.java

aip.neural.NetworkHopfield The bulk of the work in this Hopfield program is done in the *NetworkHopfield* class. This class is similar to the *Perceptron* classes, but simpler. It has two functions, training the network and executing it. Evaluating the output is the same as for the *Perceptron* classes and is done by summing the weighted inputs and applying an output function to the result. In this case, we use the *S()* sign function.

The *train()* method is unique, and implements Equation 8-35:

```
int num = m_state.getSize();
double factor = 1.0 / (double)cnt;

for (int idx=0; idx<num; idx++)
{
   double ival = template.getOrd(idx);

   for (int jdx=0; jdx<num; jdx++)
   {
      if (idx == jdx)
      { continue; }
```

```
            double jval = template.getOrd(jdx);

            double weight = m_weight[idx].getOrd(jdx) + (ival * jval *
            factor);
            m_weight[idx].setOrd(jdx, weight);
        }
    }
```

Hopfield.java

aip.app.hopfield.Hopfield The *Hopfield* application is a simplified form of the *Backprop* application. The patterns are trained into the network upon construction. Each time you select a pattern, it is fed into the network and then the network adapts to its stable position. Noise can be added to the patterns with the noise slider (Figure 8-24).

Advanced Concepts

Time-Series

The auto-associative memories explored so far are intended to stabilize on one pattern. This happens because we present the same pattern at both sides of the network. But what happens if you present different patterns to each side of the network?

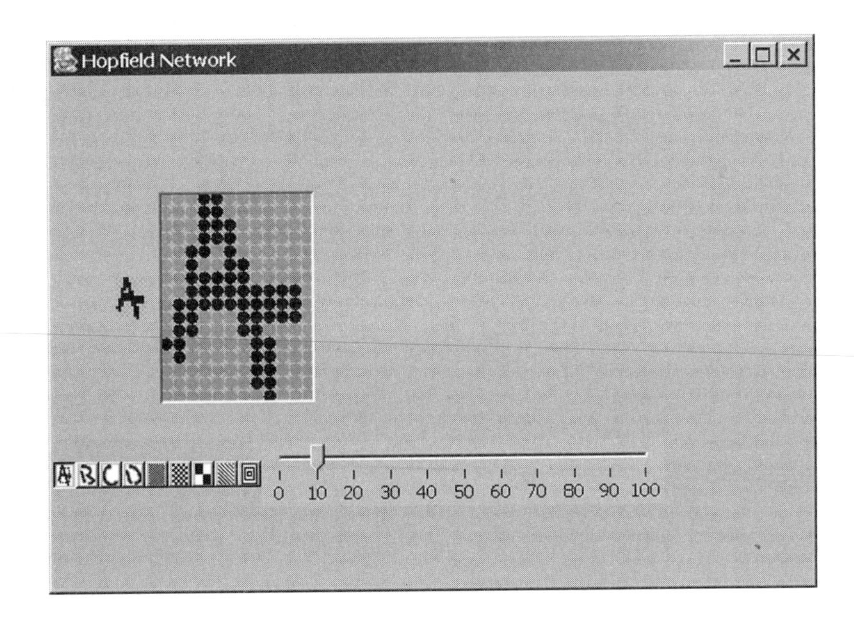

Figure 8-24 Hopfield user interface

You can test this with a simple change to the *Hopfield* program:

```
int num = m_state.getSize();
double factor = 1.0 / (double)cnt;

for (int idx=0; idx<num; idx++)
{
   double ival = template2.getOrd(idx);

   for (int jdx=0; jdx<num; jdx++)
   {
        if (idx == jdx)
        { continue; }

        double jval = template.getOrd(jdx);
        double weight = m_weight[idx].getOrd(jdx) + (ival * jval *
        factor);
        m_weight[idx].setOrd(jdx, weight);
   }
}
```

Just that one line, where *template2* is the next pattern in the sequence. With that change you can start with one pattern, clean or noisy, and the network will step through its patterns, one at a time, until it reaches the end. If you do not provide a stable point at the last pattern the network will evolve through random state space.

The backpropagation networks can also include this type of feedback, making them recurrent networks instead of feedforward. Training would be managed in the same way, but with the feedback loop turned-on during execution (Figure 8-25).

Perhaps more useful than a fully recurrent network is one that has a sense of context. Feedforward networks are context-free. They have no sense of one pattern following another. By making part of the input signal the previous output, each pattern is seen in the context of what came before it (Figure 8-26).

This context gives the network a sense of time and can help it to predict or generate time-series patterns.

Fuzzy-Neural

NN concepts can be combined with fuzzy logic to create fuzzy neural nets, giving some of the benefits of both technologies.

The fuzzy system can be attached as a front end to the neural net to generate the net's inputs from fuzzy logic statements. Or the neurodes in the network can be rebuilt so that their transfer function is a fuzzy logic statement.

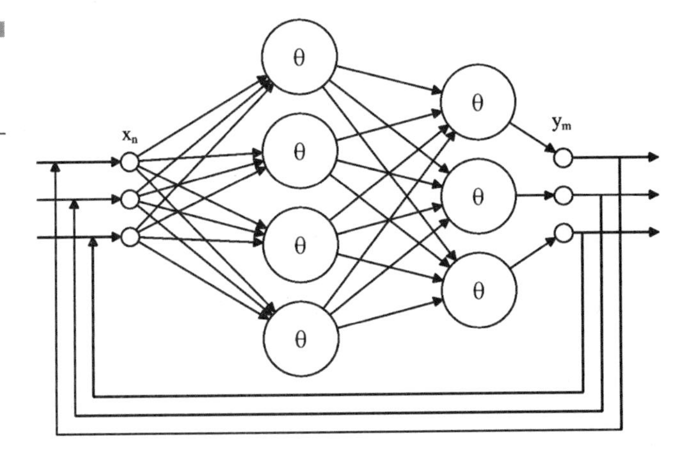

Figure 8-25
Recurrent multi-layer perceptron

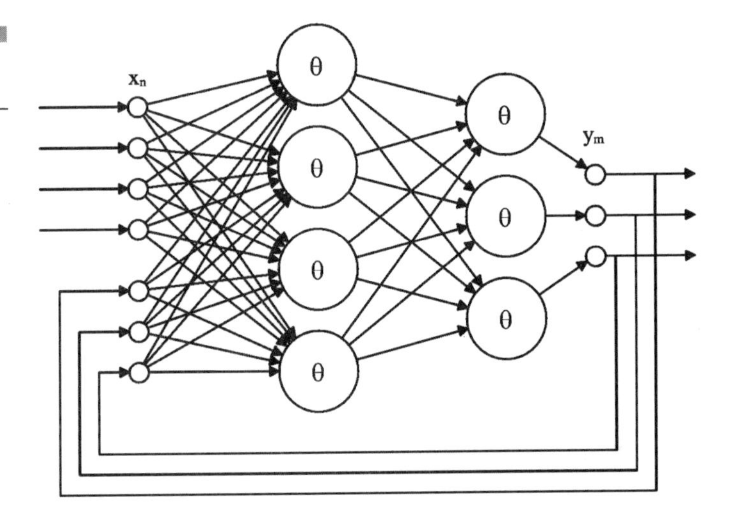

Figure 8-26
Context

There are a number of books that explore the world of fuzzy networks, three good ones being Pal (1999), Li (2001), and Kosko (1992).

Growing Networks

The exact configuration for a NN can be difficult to determine in advance. Each problem has its own set of patterns and situations, and no one network applies to all possibilities. Instead of hand-building and testing a bunch of networks, you can let the computer try them out for you.

Pairing genetic algorithms with NNs lets you evolve the optimum network for your application.

While you can use GA to evolve weights for the network, acting as a replacement for training, the more powerful method is to use GA to control the structure of the network. If you are using a three-layer network, you can specify one or more of the genes to be parameters that control, for example, the number of middle layer neurodes.

Some researchers use GA to evolve the basic architecture of the network, setting up any number of layers and even arbitrary connections between neurodes. This type of structural evolution requires a complicated coding scheme.

If you can create the network from a string of numbers, you can evolve those numbers to find an optimum network.

CHAPTER 9

Unsupervised Neural Networks

This chapter explores more details of biological neurons and computational neurons. Then we apply this information to create an unsupervised neural network, the Kohonen self-organizing map.

Once the basics are in place, we look at ways to extend the model. Some of these techniques are then applied to solve the traveling salesman problem (TSP). Finally, we look at even more extensions to the basic neural network model.

Neurons Revisited

Biological neurons inspired the field of neural networks, so it pays to have some familiarity with their operation. We can only offer the barest glimpse of the neuron here. It is a complex subject, and there are many different types of neurons and neural systems.

From these biological roots, we spend some time investigating the behavior and purpose of the computational neurons that we actually use while making our networks.

Hebb

In this section, we revisit Hebbian learning with an eye for the important concepts that it illustrates and highlights. This is just an overview, so do not worry too much about the applications and mathematical details.

Donald Hebb was trying to find the rules that drive the brain in nature. Even today neural networks are categorized by whether they are biologically plausible or not. One important attribute necessary for biological plausibility is local operation. To be biologically plausible, a neuron in the network must control its state and weights using only locally available information. This means it only gets to operate on its own state and with the state of any other neuron it actually connects with. Systems that must, for example, scan the network to locate the single most active neuron violate this rule.

The supervised neural networks violate biological plausibility in a different way as well, since they require not only the current input but the actual expected output, to be able to learn. In nature the answers are not provided, but must be discovered.

Hebbian learning, as it was initially described, is both local and plausible. The rule states that if the axon output of a neuron A is exciting a downstream neuron B at the same time that neuron B is also excited, then the efficiency of the connection between neurons A and B will be increased.

This type of learning is very Pavlovian. When two stimulations occur at the same time the association between them is increased.

The basic Hebbian learning equation can be re-written from Equation 8-35 to be:

$$\dot{w}_{ij} = y_i y_j \quad \textbf{9-1}$$

where $\dot{w}_{ij}$ is the change in efficiency, or weight, of the connection between upstream neuron i and this neuron j. y_i is the output of the upstream neuron and y_j is the output state of this neuron. Alternatively, this equation could be written as:

$$\dot{w}_{ij} = y_i s_j \quad \textbf{9-2}$$

where s_j is the internal activation level of this neuron.

Unsupervised neurons receive inputs and adjust themselves accordingly. As time progresses, new inputs and situations arise and the neurons continue to adjust. Note that the learning rule in Equation 9-1 is too strong. Each signal has too much effect on the connection weight, swamping any memory of the previous signal. To compensate for this, and to allow gradual change with time, we add the learning rate factor α:

$$\dot{w}_{ij} = \alpha y_i y_j \quad \textbf{9-3}$$

One problem with the raw Hebbian rule is that the excitation weight between neurons can only increase. If your network allows negative, inhibitory, outputs the weight can also decrease. But that still does not change the fact that the weights can increase without bound. And Hebbian neurons often only provide for excitation-based weight increases.

Though you could pass the raw weight value through one of the sigmoid functions to limit it to the unit range, that is not the traditional solution. Biological neurons exist in time and tend to forget over time. With this in mind, we can add a forgetting factor β:

$$\dot{w}_{ij} = \alpha (y_i y_j - \beta w_{ij}) \quad \textbf{9-4}$$

Now the neuron will slowly forget its associations. You can refer to the discussion on normalization for additional methods of keeping the connection weights in check. You may notice, however, that most of the normalization techniques are global in nature and violate the localness requirement for biological plausibility.

George Mobus takes the forgetting behavior one step further. In his neurons, called Adaptrodes, there are several sets of weights that operate in different time scales. One set of weights will decay very quickly, the next will decay slowly, and the last very slowly. The weights also learn at different rates, proportional to their forgetting rates.

Each weight still contributes to the total connection strength and hence behavior of the neuron.

Multiple time scales allow transitory events to have a temporary impact on the neuron's behavior without damaging its basic function. However, any stimulus that repeats regularly will eventually adjust the long-term rates and long-term behavior of the neuron.

Pattern Matching

Starting with the McCulloch-Pitts neuron and proceeding through the multi-layer perceptron to the associative memories, we have seen neurons and neural networks acting as pattern matching devices.

Networks of neurons are also pattern generation devices, with each neuron creating a single bit or value within the pattern.

The neurons used so far perform their input pattern match using the dot product of the input vector against the weight vector. For normalized inputs and weights, this returns a value from –1 to +1 that is the cosine of the angle between the vectors.

Another approach is to take the Euclidian distance of the input vector to the weight vector. This form is used in radial basis function (RBF) networks, which are similar to the multi-layer perceptron of the previous chapter. In this case, though, the weight is considered an exemplar, reference vector, or codebook vector that defines a position in state space.

Figure 9-1 shows the dot-product response curve, as well as two forms of RBF response. The RBF equation can take two forms, as shown in Equation 9-5. The form with the square root gives a sharper response.

$$y = \exp\left(-\sum (c_i - x_i)^2\right)$$
$$y = \exp\left(-\sqrt{\sum (c_i - x_i)^2}\right) \qquad \text{9-5}$$

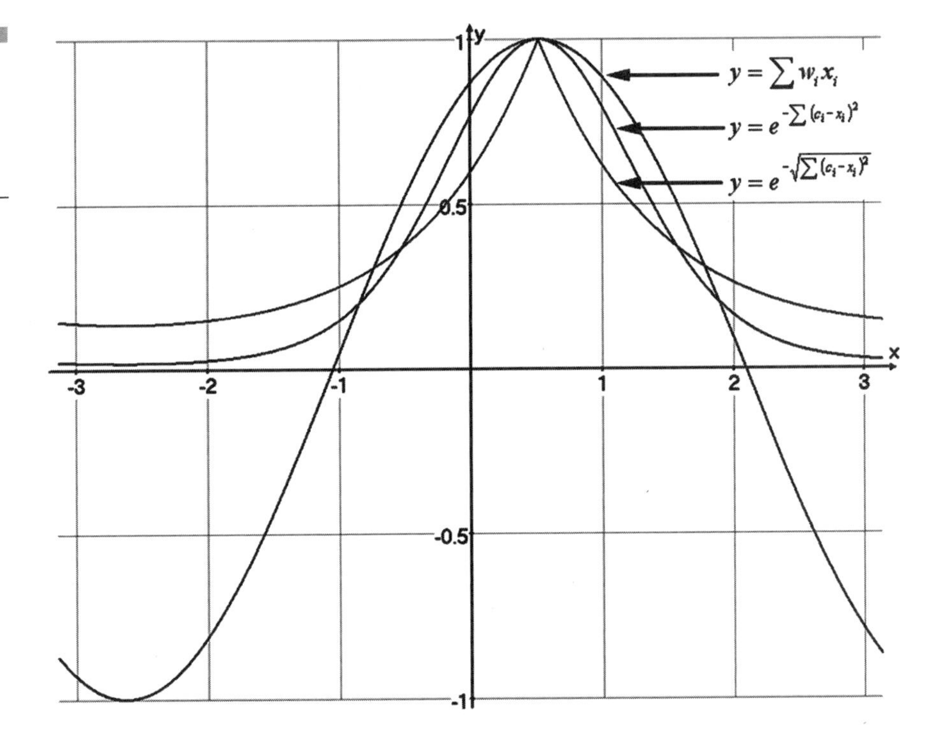

Figure 9-1 Dot product vs Euclidian activation functions

The RBF response can be sharpened even further using a control factor λ. Larger values of λ make the response more focused.

$$y = \exp\left(-\lambda\left(\sqrt{\sum (c_i - x_i)^2}\right)\right) \qquad \textbf{9-6}$$

On the output side, each neurode can be considered to have a large number of axons that lead downstream. These outputs can be weighted so that when a neurode becomes active, it sends out another pattern, whatever was imprinted in the output weights. Note that in the MLP network, the weights on a neurode's output layer and the weights on the downstream input are the same weights.

Stephen Grossberg developed this fresh way of thinking about neurodes. A pattern matching neurode is called an instar in his system, since it can be drawn like a star with data streaming into it. A pattern generating output neuron is an outstar (Figure 9-2). Though Grossberg applies these concepts to different types of networks than those we explore here, the separation of input pattern matching from output pattern generation can help you visualize the operation of any network.

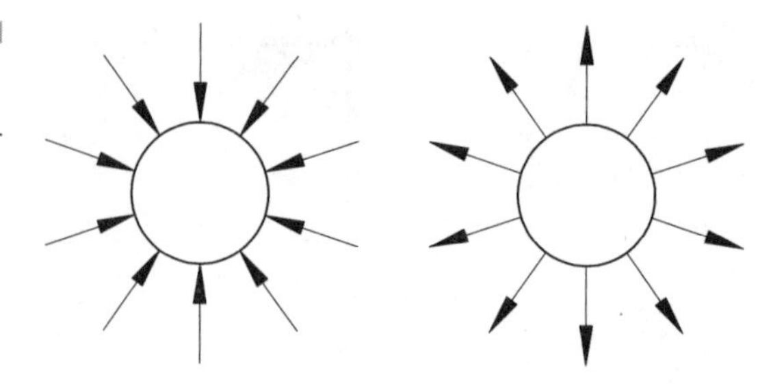

Figure 9-2 Instar and outstar

Self-Organizing Maps

All of the supervised networks in Chapter 8 were trained by giving them an input pattern and an output pattern together. These input/output pairs are used to force the weights into forms that convert the given input to the given output. This is fine if your problem has well-defined inputs and outputs that you can train against, but this is not always the case.

Perhaps you do not know the precise conditions or areas of state space the network must respond to, or perhaps the conditions change with time. Regardless of the reason, some applications need the network to learn without being spoon-fed the answers. These self-training networks use unsupervised learning, and an important form of unsupervised learning is found in self-organizing maps.

Teuvo Kohonen pioneered the self-organizing map (SOM) in 1982 (Kohonen, 2001), and today you can find thousands of research papers on the subject. Another excellent SOM book is Obermayer (2001). SOMs learn to categorize input vectors using nothing more than the statistical distribution of those inputs.

Data from a continuous process, or sensory input from the outside world, tends to fall into a limited area of state space. Each neurode in a SOM stakes out its own piece of state space, with more neurodes clustered in heavily traveled areas.

Each neurode in the SOM is a pattern matcher, as described above, using the Euclidian, or more biologically plausible dot-product, vector match. These neurodes are organized as a one-dimensional line or circle, a two-dimensional square or hexagonal grid, or any other N-dimensional network.

When an input is presented to the SOM, each neurode tests its reference vector against it. The neurode with the best match, and its close neighbors in the SOM network, are then adjusted to better match the input.

Over time, the neurodes organize their reference vectors in state space. Each neurode only responds to inputs near its area of state space, categorizing that input as a position in the SOM network.

This input categorization reduces the degrees of freedom of the input down to the dimension of the SOM network. For example, the input vector could be composed of 10 different ordinates, but the SOM network might be a two-dimensional array. The 10 dimension input is projected onto the two-dimensional SOM.

The SOM preserves the relationships of the input data as well, so similar inputs map to neurodes that are near each other in the network. The SOM makes an excellent front-end for other applications or AI systems since it reduces the dimensionality of the input while preserving most of its meaning, compressing it into a smaller state space. For example, an SOM makes a good preprocessor for reinforcement learning.

Now we can look at the three aspects of the SOM in detail, starting with the map architecture.

Basic Operation

In the MLP, the neurons sent their results downstream to the next set of neurons, and so on until an output was generated. SOMs are one-layer categorizers and as such do not have a "downstream". In this respect they are much like the simple perceptron. What they do have, however, is the concept of neighborhood.

Each neuron in the map is connected to a number of neighbors. How many neighbors depends on the dimension of the map. Figure 9-3 shows some one-, two-, and three-dimensional map structures. There is no theoretical limit to the map's dimension, though one- and two-dimensional maps are preferred, because they are easy to draw.

Hexagonal networks are preferred over square networks because square nets tend to limit themselves to simple vertical and horizontal patterns (Kohonen, 2001). Square networks, however, are usually easier to draw. Interestingly enough, rectangular arrays can provide more stable results than square or circular layouts.

There are two ways to define the neighborhood of an active neurode. The simplest method is to use the connectivity of the neurode graph. The first set of neighbors are all those neurodes directly connected to the

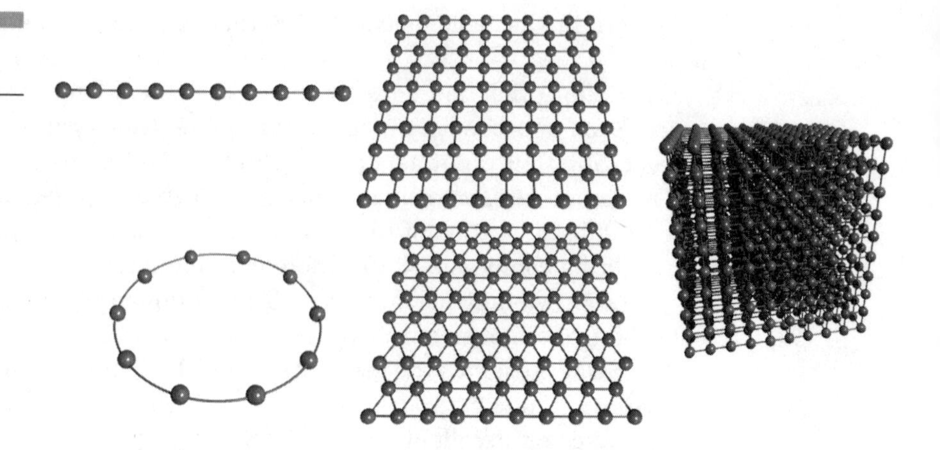

Figure 9-3 SOM lattices

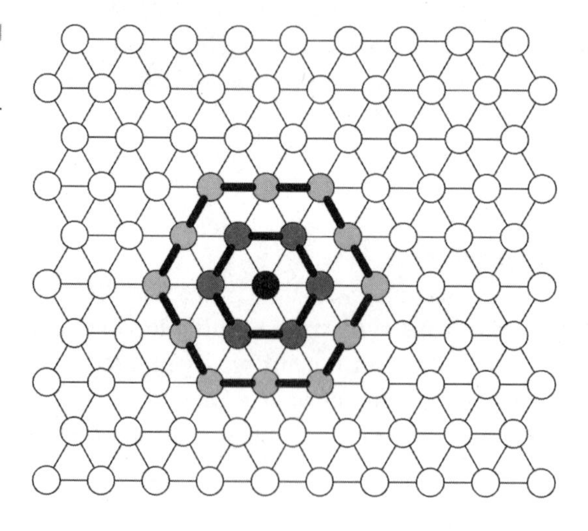

Figure 9-4 Neighbors

active neurode. The second set of neighbors are the neurodes connected to the first set, and so forth (Figure 9-4).

Neighbors can be identified by distance as well, which is subtly different than the connectivity neighborhood described above. For this to work, the neurode map must be considered to be embedded in space and not just an abstract set of connections. Each neurode then has a coordinate in space and it is then a simple matter to calculate the distance between any two neurodes.

The actual neurode coordinates and distances are arbitrary and will depend on the layout of your map.

The network can be initialized so that each neurode's reference vector is random, though training proceeds faster if the vectors are spread evenly throughout state space.

Every input is sent to every neurode in the map. As the SOM increases in complexity, this rapidly becomes impossible to draw (Figure 9-5), so we tend to cheat and show the map edge-on or with some other simplification.

As usual, each neurode in the SOM becomes activated to the extent that its reference vector matches the input vector.

Once the input has been processed by the neurodes, external program code can scan the network and find the best-matching neurode. For Euclidian tests this will be the neurode with the smallest value, and for dot-product tests it will have the largest value. The position of this neurode in the map can then be used as the category identifier for the input.

The use of external code to determine the winning neurode clearly breaks any connection with biological plausibility. We address this problem later.

In the beginning, the category matches will not be very good so the network must be trained. Training and matching can operate together so the network can adapt to changes in the input space.

Once the winning neurode has been selected, its reference vector is adjusted to be closer to the input vector. Neighboring neurodes are also moved towards the input, but to a lesser extent:

$$h_i = \exp\left(-\frac{d^2}{\sigma^2}\right)$$
$$\dot{r}_i = \alpha h_i (x - r_i) \tag{9-7}$$

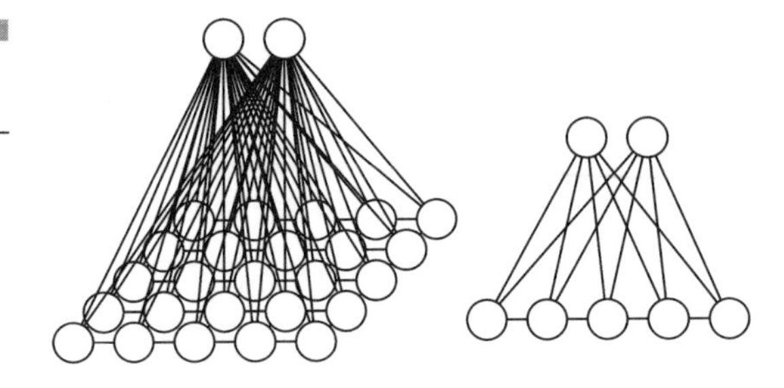

Figure 9-5
SOM input

where:

h_i is the neighborhood factor, described in the text
d is the distance of this node from the winning node
σ determines the width of the neighborhood
r_i is the reference vector
x is the input vector
$\dot{r}_i$ is the change in the reference vector
α is the learning factor

How much the reference vector for a neurode changes depends on how far that neurode is from the winning neurode. Note that the distance d is in terms of distance on the map, and not the distance between the reference vectors. The range of influence is modified by σ. The neighborhood function in Equation 9-7, using sample σ values of three, two, one, and one-half, is shown in Figure 9-6. There are also other neighborhood functions that you could use.

As training progresses, the neighborhood width factor σ is slowly reduced to its lower limit. The learning factor also starts out relatively large and reduces to its minimum. At first, each input vector has a broad ordering influence on the map, causing many, if not all, of the reference vectors to adjust. As the neighborhood width is reduced, the neurode adjustments become more and more focused, until the map stabilizes.

Sometimes your test data will not be spread evenly across state space. If this is the case, you can enhance the impact of important, yet rare, test examples by increasing α for them. Similarly, you can force a pattern into

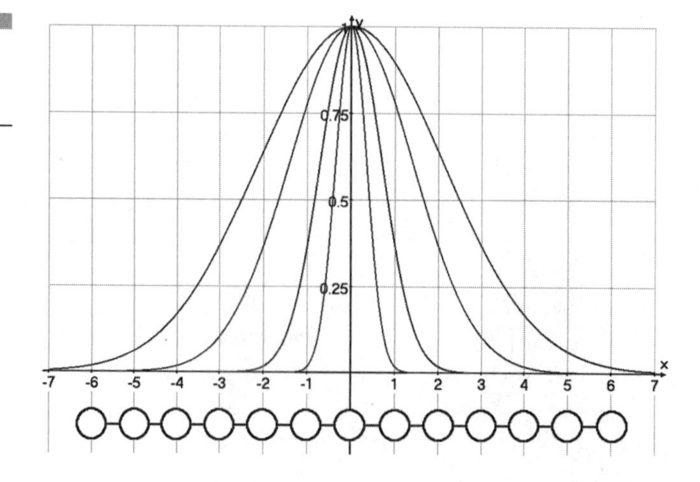

Figure 9-6 Neighborhood function

a fixed area of the map by initializing the neurode or neurodes in that area with the sample vector, and then decreasing α when these preset patterns appear during training (Kohonen, 2001).

Note that we are operating in two separate spaces. The reference vectors and the input vectors are all from the application's state space. The neighborhood function for the vector update is in the SOM network space. The input space is projected onto the network space.

Code: RGB Map

This test program maps the three-dimensional red/green/blue color space down onto a two-dimensional square-grid SOM. This provides a good, visceral test of the algorithm, watching the colors organize themselves into clusters during the test runs.

SOMs have a difficult task when it comes to displaying the state of the network. There are three types of information stored within the SOM.

The first obvious information is the activation level of the neurodes, which is what we displayed in the previous networks. However, in the SOM model we often want to display the reference vectors. This can be complicated since these vectors can be large. Finally, the structure of the SOM is significant, since each neurode affects its neighbors.

The *SOMColor* test does not display the activation level but instead displays the reference vectors of the neurodes. These are conveniently mapped into the RGB color of each neurode's display area. The structure of the SOM network is implied by the position of the neurodes in the display.

Other types of SOM networks may want to display their reference vectors differently, such as with a little graph inside the neurode, or some other application-specific representation. If the input vector is of a sufficiently low dimension, it can be used as a coordinate on a graph, where a dot can be drawn. The network neighborhood can then be drawn as lines between the nodes on this graph.

The reference vectors do not all have to be drawn into one display, either. If there are related subsets in the input vector, each subset could be drawn into its own graph. For the color example, there could be separate displays for the red, green, and blue components.

The class diagram for *SOMColor* is shown in Figure 9-7.

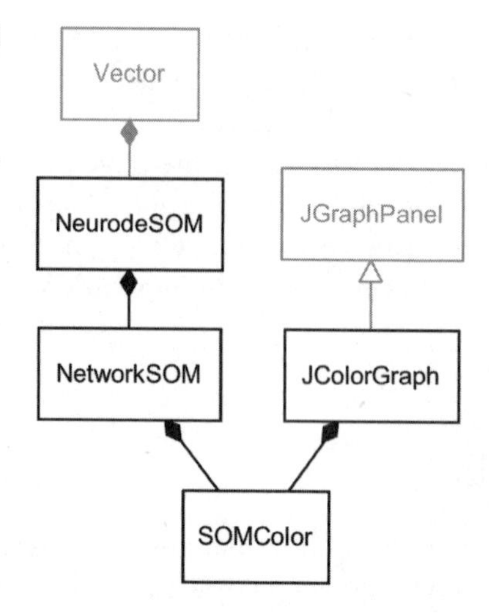

Figure 9-7 SOMColor class diagram

Table 9-1 JColorGraph summary

Construction	
	`JColorGraph (int dx, int dy)`
`void`	`create ()`
Data Feed	
`int`	`addLayer (String label, Color color)`
`void`	`setRange (int dim, double min, double max)`
`int`	`getDimension ()`
`void`	`addDatum (double x, double y, Vector color)`
`void`	`done ()`

JColorGraph.java

aip.display.JColorGraph This is a specialized version of *JGraphPanel* that displays a grid of colors. The data is presented to it as a three-dimensional *Vector*. A summary of *JColorGraph* is given in Table 9-1.

NeurodeSOM.java

aip.neural.NeurodeSOM The *NeurodeSOM* class provides the testing and training behavior for the SOM. The state test can be performed with the dot product with *testDot()* or with Euclidian distance using *testDist()*. Otherwise this is a very simple class whose job is to make *NetworkSOM's* life easier.

A summary of *NeurodeSOM* is given in Table 9-2.

The *testDot()* method is just an alias for the *Vector* dot product:

```
m_state = input.dot(m_ref);
```

Table 9-2

NeurodeSOM summary

```
Construction
                NeurodeSOM (int size, Vector pos)
          void  reset ()

Operations
        double  testDot (Vector input)
        double  testDist (Vector input)
          void  train (Vector input, double step)

Access
           int  getSize ()
        Vector  getReference ()
        double  getState ()
        Vector  getPos ()
```

While *testDist()* is a distance calculation:

```
Vector delta = new Vector(m_state);
delta.sub(input);
m_state = delta.length();
```

Training with *train()* moves the reference vector closer to the input by some step amount:

```
Vector delta = new Vector(input);
delta.sub(m_ref);
delta.scale(step);

m_ref.add(delta);
```

NetworkSOM.java

aip.neural.NetworkSOM The network of SOM neurodes processes input patterns and performs the global test for the best-matching neurode. Once the input has been spread around and the neurodes all have a current active state, *NetworkSOM* makes a second pass across the neighbors of the winning neurode and trains them according to the current learning and neighborhood factors.

A summary of *NetworkSOM* is given in Table 9-3.

NetworkSOM(int num, Vector input, boolean wrap) The constructor sets up the network and fills it with random reference vectors. The *num* parameter indicates the size of the network, where there are a total number num^2 neurodes arranged in a square. The *input* vector is used to establish the size of the input patterns, while *wrap* controls whether the neurodes at the edge of the map interact with the neurodes on the opposite side or not.

Table 9-3

NetworkSOM summary

Construction	
	NetworkSOM (int num, Vector input, boolean wrap)
void	reset ()
Operations	
Vector	stepMax (Vector input)
Access	
double	getLearnFactor ()
void	setLearnFactor (double factor)
void	scaleLearnFactor (double scale, double min)
double	getNeighborhood ()
void	setNeighborhood (double size)
void	scaleNeighborhood (double scale, double min)
NeurodeSOM[][]	getNeurode ()
NeurodeSOM	getNeurode (int x, int y) throws IllegalAccessException
int	getSize ()

reset( ) This method resets the reference vectors to random values.

getLearnFactor( )
setLearnFactor(double factor)
scaleLearnFactor(double scale, double min) These methods retrieve and specify the learning factor for the network. The *scaleLearnFactor()* multiplies the current learning factor by the *scale* value, which is typically just a shade smaller than one. During scaling, the learning factor is not allowed to shrink below *min*.

getNeighborhood()
setNeighborhood(double size)
scaleNeighborhood(double scale, double min) Get, set, and scale the neighborhood size. A neighborhood value of one covers roughly one neighbor around the center.

SOMColor.java

aip.app.somcolor.SOMColor The *SOMColor* test application provides an interface to control and view the *NetworkSOM* object. This is a simple interface, as shown in Figure 9-8.

The bulk of the window is taken up by the color dots that display the reference vectors. To the right of this are the controls.

The three options *5 Colors*, *8 Colors*, and *All Colors* determine the input state space. The first two limit the choices to fully saturated colors,

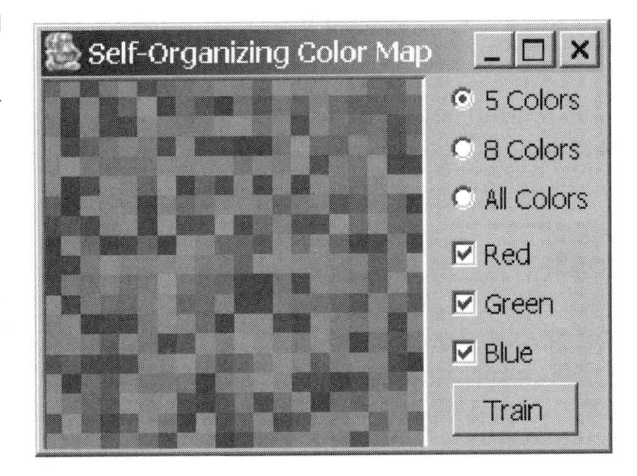

Figure 9-8
SOMColor UI

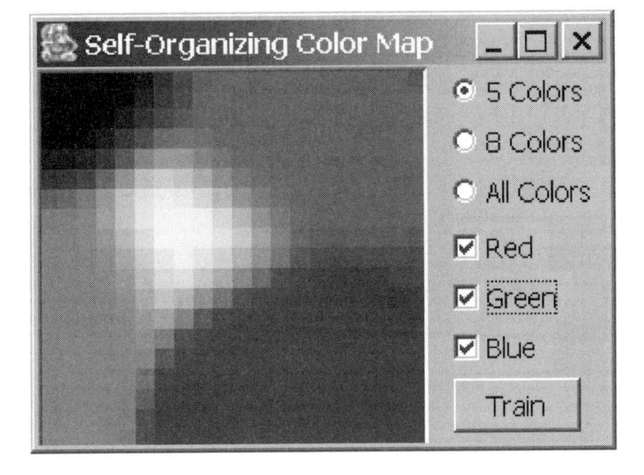

Figure 9-9
Organized SOM

as defined in the array *m_color_table*. The *All Colors* option selects input colors from the entire RGB space, which tends to overwhelm the map.

The three color checkboxes, *Red, Green,* and *Blue* only affect the display. Turning these on and off lets you see the organization of the individual components of the reference vectors. On a well-organized map, the red values will be clustered together and the same with green and blue. Of course, sometimes it splits colors, especially when many colors are in use.

The *Train* button randomizes the network and starts the training. Once the network settles, you will see a map like Figure 9-9, which shows the result of training in the five color state space.

Network Variations

SOMs, and neural networks in general, come in many different variations. Here are a few variations on the SOM theme, some of which are presented in detail.

Supervised

The unsupervised learning algorithm can be adapted to a supervised training form, which can be useful if you want more control over the organization of the map. This is done by training the SOM with extra category information, and then executing without these categories. This is illustrated in Figure 9-10.

When the SOM is being trained, its input pattern consists of two parts. The first piece is the raw data. This is the type of data that it is expected to categorize during execution. The second piece is a pre-calculated, or pre-defined, category for this data. For example, for speech processing the raw data could be a digitized segment of speech, and the category could be its phonetic category (Kohonen, 2001).

Once trained, only the raw input is presented to and processed by the SOM.

Pre-initializing the reference vectors to evenly cover the expected state space can accelerate training. Random initialization is used to demonstrate the power of the system, but is not necessary or even desirable in practical use.

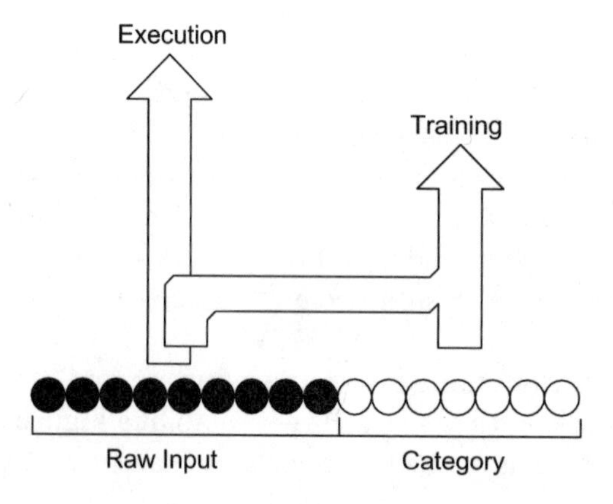

Figure 9-10 Supervised SOM input

Biologically Plausibility

The basic SOM model requires a global observer to note which neurode is closest to the input pattern. This is not a reasonable strategy from a biological point of view.

If we relax the requirement of having only a single winning neurode, we can use the mechanism of lateral inhibition to focus the network's response. Each neurode changes its neighbor's activation level, stimulating nearby neighbors while inhibiting distant ones, through the "Mexican hat" distance function (Figure 9-11):

$$h(x) = (1 - x^2)\exp\left(-\frac{x^2}{2}\right) \quad \textbf{9-8}$$

You can scale x, before its use in the $h(x)$ calculation, to control the distance d where you want the function to cross zero:

$$x = \frac{x}{d} \quad \textbf{9-9}$$

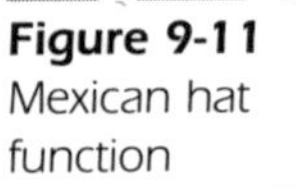

Figure 9-11 Mexican hat function

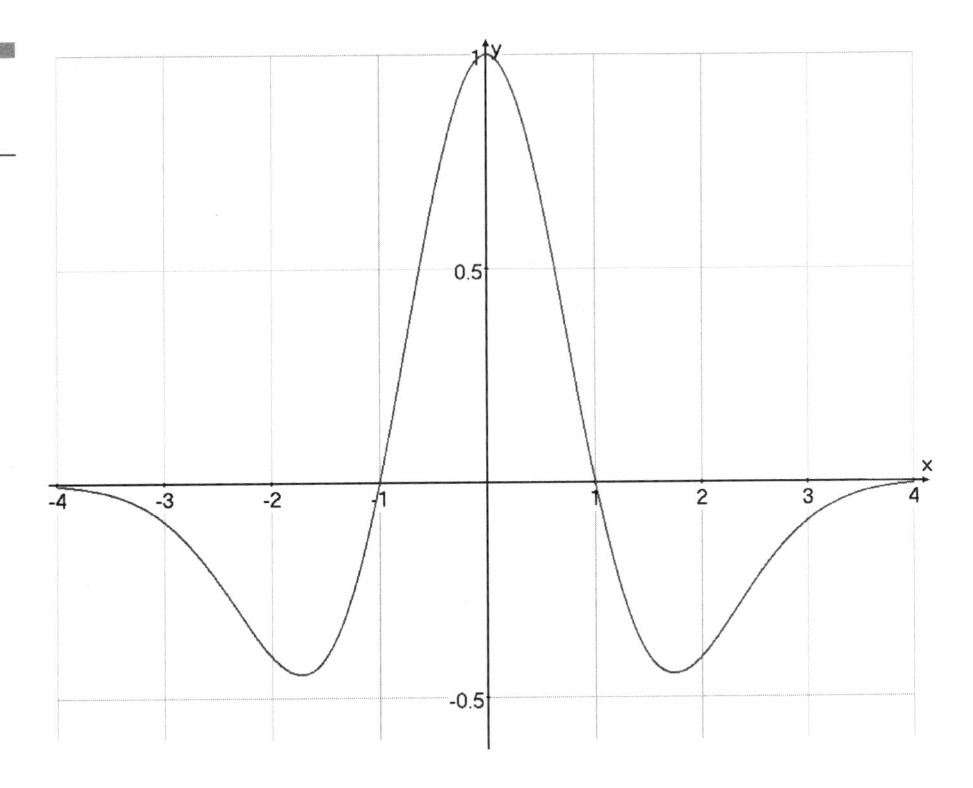

The Mexican hat is one of the RBFs, and can be used as an alternate activation function for some backpropagation networks.

The total adjustment, taken across the entire range of the function, is zero. The effect is to sharpen the edges, or differences in activation.

The new state of a neurode is based on its match with the input pattern plus any contributions based on the distance-adjusted previous state of its neighbors:

$$s'_{ij} = \sum_k w_{ij,k} x_k + \sum_{u,v} h(uv,ij) s_{uv} \qquad \textbf{9-10}$$

where $h(uv,ij)$ is a function of the distance between neurode i,j and u,v.

Once a pattern has been presented to the network, the initial neurode states are set. The network is then iterated a number of times with this input so that the neighborhood adjustments can settle out.

The details behind this form of SOM can be found in section 7.2 of Miikkulainen (1993).

Related to this is Risto Miikkulainen's trace feature maps, also in Miikkulainen (1993). In these networks, the neurodes not only learn their place in the input state space, but adjust the connection weights to their neighbors. These weights form a "trace" in state space that can funnel the network activation over to the best matching neurode.

Growing Networks

One of the difficulties with neural network design is finding the right number of neurodes for the network. While we briefly looked at genetic algorithms to evolve networks, SOMs have a different technique available to them.

As patterns are entered into the SOM, there will be one winning neurode that responds the strongest to a pattern. If the inputs are clustered in one area of state space, the neurodes in that area will be activated more frequently than neurodes in sparsely used areas.

If the neurodes are given the ability to track how often they are used, frequently used neurodes could split up and cover the dense state space with a finer grain of control, and relatively unused neurodes could be deleted.

Each time a neurode c wins in the SOM, its counter τ is incremented by one, while all other neurodes have their counter decreased:

$$\begin{aligned}\dot{\tau}_i &= \delta\tau_i \\ \dot{\tau}_c &= 1\end{aligned} \qquad \textbf{9-11}$$

where $\dot{\tau}$ is the change to apply to the neurode, and δ is the forgetting factor.

The activation level can then be used directly from the value of τ, or you can calculate the neurode's relative activation level:

$$h = \frac{\tau}{\sum_i t_i} \qquad \textbf{9-12}$$

Once the system decides a neurode is overworked, a new one is inserted next to it. If a neurode is being under-utilized it is removed.

Flexible Lattice

If we are going to be adding and removing neurodes from our map, we cannot think of it as a rigidly defined line, grid, or other N-dimensional array. We need to have a flexible lattice. The simplest form is based on the triangle, tetrahedron, or other k-dimensional structure from that family.

This class of lattice has $(k + 1)$ neurodes at the vertices and $(k(k + 1))/2$ edges, or connections on each neurode. We will use the triangular lattice of dimension $k = 2$ for our growth experiments, as shown in Figure 9-4.

We need to be able to both insert and delete nodes in this lattice. Note that while we show it all nice and tidy in our current figures, inserting and deleting nodes changes the density of the lattice near those modifications.

These techniques of SOM lattice modification are adapted from Fritzke (1993) and are also explored in Wise (2003).

Insert a Node

New nodes are inserted between two existing nodes, splitting their connection. The new node has the first and second nodes as neighbors. It also has as neighbors the shared neighbors of the original nodes. This is shown in Figure 9-12, with the gray new node inserted between the two black nodes.

We already looked at the criteria for when to insert a new neurode; when one of the existing neurodes has become too popular and needs a

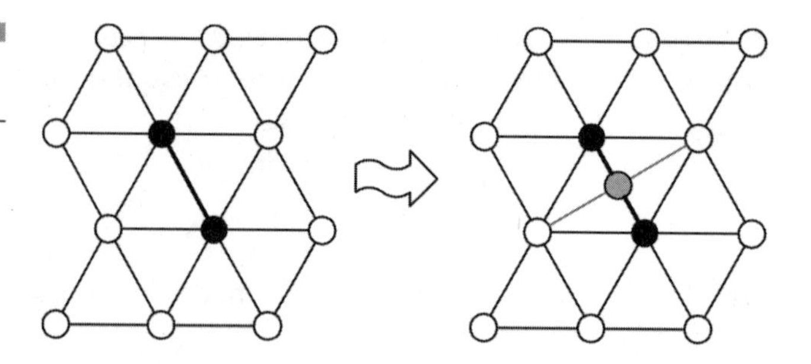

Figure 9-12
Inserting a node

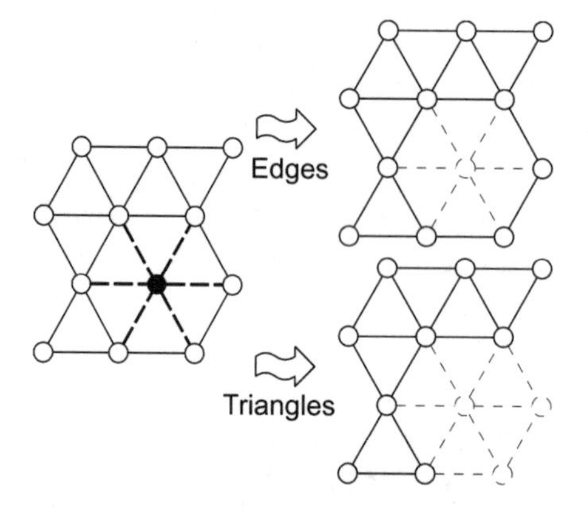

Figure 9-13
Deleting a node

clone to keep up with its duties. But how do we select the second node in our pair?

The second node is the neighbor of the overworked node that is the farthest from that node in state space. That is, its reference vector is the most different.

Once the new node is created, we need to assign it a counter value and a reference vector. The new reference vector can be placed halfway between the two parent nodes in state space. The new counter can take points from the overworked parent node, so they each have about half of the original count.

The exact system for allocating counter values and placing reference vectors is not too critical. These values adjust as the network processes new inputs, until they match the statistical properties of those inputs.

Delete a Node

If our example neurode were slated for removal instead of splitting, the process looks like one of the examples in Figure 9-13.

The first, obvious, method of deleting a neurode from the lattice is to simply remove it and all of its edges. This, however, can leave edges floating around in violation of our structure, as shown in the "edges" result in Figure 9-13.

The correct method is to look at all of the triangles that share that node. Each triangle's edge must share an edge with another triangle for it to be preserved. Any triangle without any anchored edges is deleted.

Note that deleting a node can split the network into two or more independent sections! This is allowed, though there is no provision for merging these sections back together. The counter information for the deleted node can be shared out among its neighbors or simply thrown away, as the occasion warrants.

Fritzke (1993) goes into more detail on the adding and deleting process, though in practice I have found it is not necessary to get too fancy. Additional references on SOM growth can be found in Oja (1999) and Obermayer (2001).

Growing Grids

While you need a lattice as described above for single-neurode insertion, there are times when you want to keep a regular grid structure for your SOM. In this case, once you have identified the node to split and its farthest neighbor in state space, you insert an entire column or row of neurodes instead of just one. This stretches the grid by one column (or row). Each neurode inserted in that line takes on a reference vector that is midway between its neighbors.

A more detailed exploration of growing grids can be found in Fritzke (1995).

Elastic Networks for the Traveling Salesman Problem

The elastic net (EN) is a variation on the Kohonen self-organizing map. It is particularly good when applied to the TSP. Our implementation of the EN incorporates the growing network logic outlined above. But first an explanation of the TSP itself.

The TSP is easy to define and difficult to solve. In it, you have a large number of "cities" (Figure 9-14), or more likely holes in a circuit board or some other manufacturing problem, and you want to find the shortest path that visits each city once and only once.

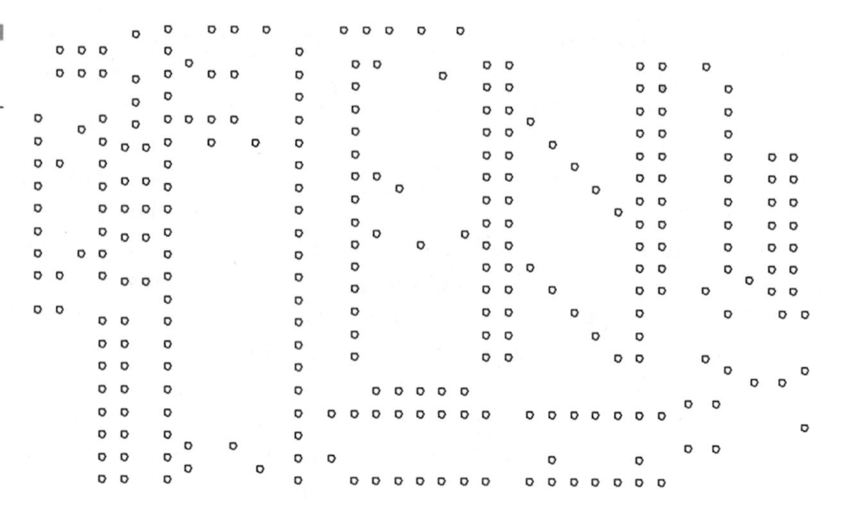

Figure 9-14
"a280" city layout

We mentioned the TSP in Chapter 6 as an example for permutation coding. Here we solve the TSP using the SOM-like elastic network.

The brute-force approach to the TSP is to check every possible path through the cities and then choose the shortest. For any reasonably sized problem, this is effectively impossible.

There are many different approaches to solving the TSP and some of them are very good, better than the EN. A list of many of these solutions and their relative merits can be found, at the time of this writing, at `http://www.research.att.com/~dsj/chtsp/results.html`.

The EN solution is presented here because of its association with SOM techniques. The EN was introduced in 1987 by Richard Drubin and David Willshaw in *Nature* magazine (Durbin, 1987). It is simple and surprisingly effective.

The environment the EN operates in consists of the cities (Figure 9-14) and a path defined by the nodes in a one-dimensional network. At each step of operation, the nodes are attracted to the cities and a surface tension is applied, where each node is attracted to its neighbors. The path is like a rubber-band and the nodes and cities are like magnets in this band that are drawn to each other. The attractive forces are narrowed over time as the nodes draw closer to the cities, until the path reaches every city.

The standard EN algorithm requires many more nodes than cities. It provides a smooth and efficient solution to the TSP, reaching path lengths that are within a few percent of perfection.

This version of the EN starts with just a few nodes and adds more where the cities are dense. We take other liberties with the algorithm as well, to reduce the number of calculations and iterations needed to reach the final path. A side effect of these shortcuts, however, is a reduction in the quality of the solution.

In research, the primary goal is to find a perfect solution and performance is a secondary, though still important, consideration. The environment that I developed this EN variation in is quite different. My manufacturing clients are typically not worried about the optimum path, but they are very interested in calculating the path quickly. Not in minutes, but in seconds or even better, milliseconds. To achieve this goal, I need to find ways to cut out all extra operations, including the traditional Gaussian distance function.

Another interesting constraint I face reduces efficiency even further. Not only do my clients want a fast path through their machining, they want it to "look right". Instead of the round-about, looping, squiggly path that may be the most efficient, they would rather have a tidy zig-zag path that is predictable and orderly. In some cases, the machine requires this kind of zig-zag path or else it can be damaged! In other cases, if the machine does something unpredictable, the operator can be damaged.

Fast Elastic Net As mentioned above, the EN environment consists of a number of cities and the nodes in a one-dimensional path. When the network is initialized, a small number of nodes are created in a loop as shown in Figure 9-15.

If you want a different final path shape, you create different starting paths, such as a zig-zag or box. In one application, I create a zig-zag path

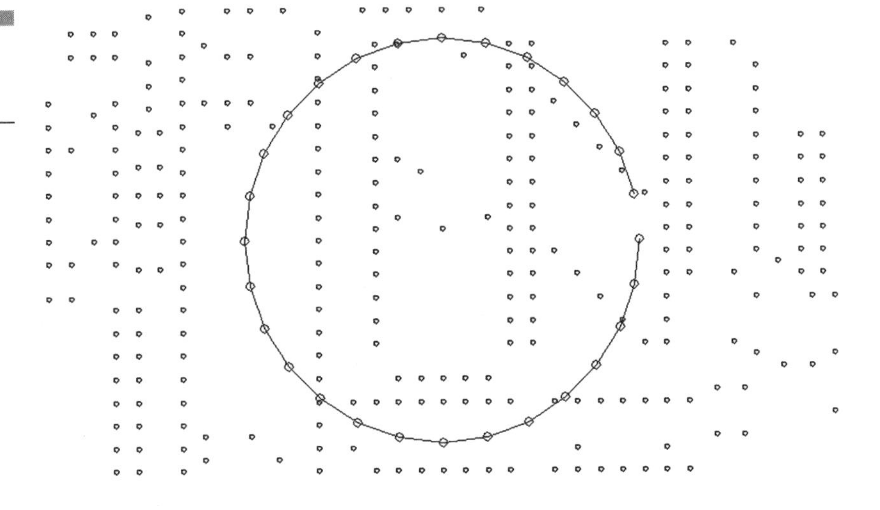

Figure 9-15 "a280" starting path

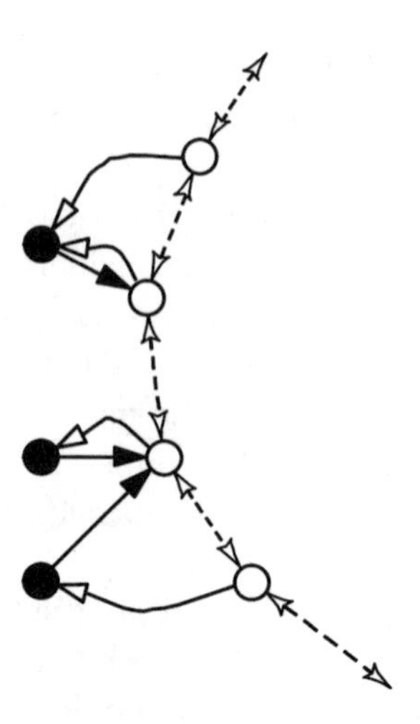

Figure 9-16
Cities and nodes

that extends outside of the range of the cities. I then lock down the nodes at the end of each zig and zag. Once the network has been solved, it follows a tidy back and forth path through the cities.

This example, however, is limited to a circle where all of the nodes are mobile.

On each cycle, each city is associated with the nearest node, and each node is associated with the nearest city, as shown in Figure 9-16.

Note that these associations are not necessarily symmetrical. The city a node chooses may not choose that node back, and several nodes may select one city while several cities can select one node.

Once the associations have been made, each city exerts a pull on its selected node, and each node is pulled towards its selected city. These pulls are then balanced by the surface tension between neighboring nodes.

Once all of the forces have been calculated the nodes are moved and the cycle repeats.

Since we start with fewer nodes than cities, the first cycles calculate quickly, developing a rough outline of the path.

When multiple cities select a single node, the weight counter for that node increases disproportionately. When it grows large enough the node is split into two nodes, both of which are near the original position.

If a node is not getting selected much at all, it eventually dies.

Once each city has a node on top of it, within a specified tolerance, and there are no spare nodes floating around between cities, the path is complete.

One problem with the EN solution is that it tends to develop kinks in its path, like the one in Figure 9-17.

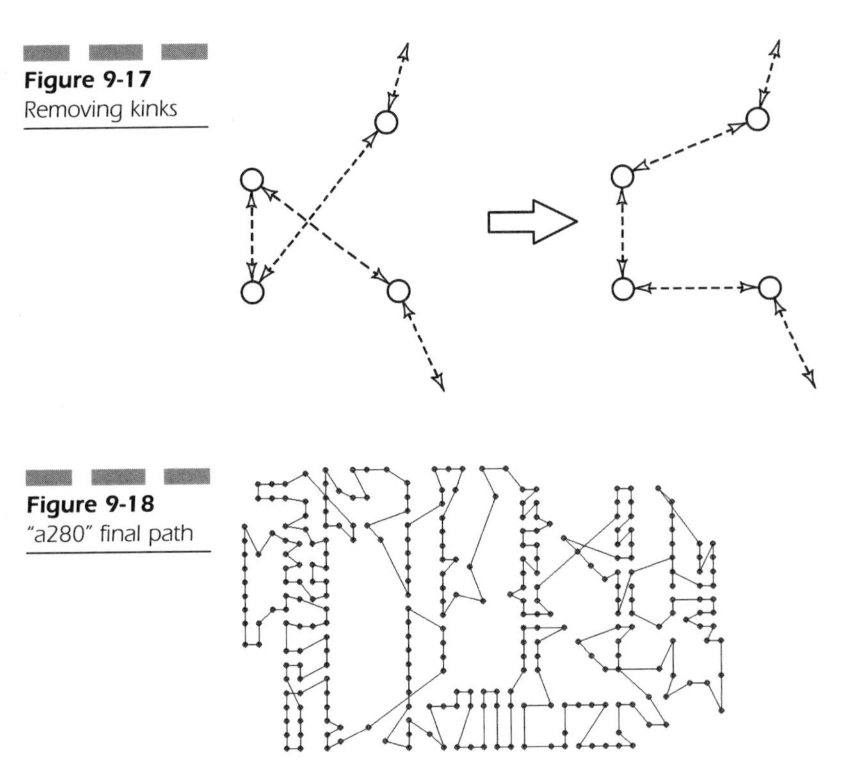

Figure 9-17
Removing kinks

Figure 9-18
"a280" final path

A final pass through the path can remove these kinks. This final "greedy swap" pass swaps the position of two nodes to see if this makes the path any shorter. If it does, it keeps the swap, otherwise it tests more nodes.

The final result is an efficient path that visits all of the cities, as shown in Figure 9-18. This path is not perfect, in fact it could be considered terrible, about 20% longer than optimal! But it also calculates in an average of about 40 generations, and many of those generations have less nodes than cities. We have traded efficiency for blinding speed. And, with starting path variations, we can have total control of the path's shape.

Code: Fast Elastic Net

Most of the code in this TSP application is involved with reading the city layout from an external file and drawing the system state to the screen. The EN calculations are all performed in the *NetworkElastic* class, which includes not only the elastic network but also the greedy swap cleanup.

Figure 9-19
TSP class diagram

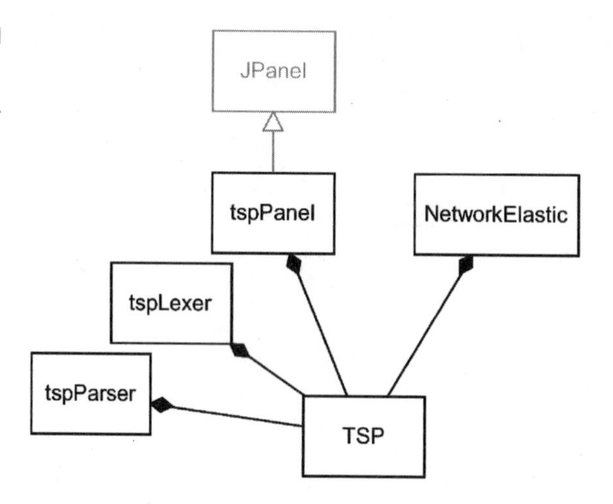

The class diagram for TSP is given in Figure 9-19.

The test data used in these examples comes from TSPLIB at `www.iwr.uni-heidelberg.de/groups/comopt/software/TSPLIB95/`.

tspParser.java
tspLexer.java
aip.app.tsp.tspParser
aip.app.tsp.tspLexer These classes provide an interpreter for the *.tsp* city data files. Instead of creating a state machine to parse the files, we are using a version of the classic Lex/Yacc programs from Bumblebee Software (`www.bumblebeesoftware.com/`). Their *Parser Generator* package can be found in its entirety under the aip/app/tsp package.

We will not spend any time looking at the parser here. You can find any number of good references for Lex and Yacc, such as Levine (1995).

These java files were generated by *Parser Generator* using the templates *tspLexer.l* and *tspParser.y,* which are part of the TSP package.

tspPanel.java
aip.app.tsp.tspPanel The *tspPanel* display is loosely based on the *JGraphPanel*-based classes from the *aip.display* package. However, it is specialized for the display of the cities and nodes in the TSP problem and it only has a few public methods. A summary of *tspPanel* is given in Table 9-4.

tspPanel(int dx, int dy) The constructor sets up the tsp display panel at a given size, in pixels.

Table 9-4

tspPanel summary

```
Construction
                tspPanel (int dx, int dy)

Drawing Operations
          void  clear ()
          void  drawDot (int style, double x, double y)
```

Table 9-5

NetworkElastic summary

```
Construction
                NetworkElastic ()
          void  addCity (double x, double y)
          void  reset ()

Operations
       boolean  step ()

Access
        double  pathLength ()
           int  generation ()
      Iterator  cityIterator ()
      Iterator  nodeIterator ()
```

clear() Erases the contents of the panel.

drawDot(int style, double x, double y) Places a dot into the panel. Note that the TSP system scales its data into the (*X*, *Y*) range of (–1.0, –1.0) to (+1.0, +1.0). As such; the *x* and *y* ordinates passed to *drawDot()* are also in that range.

The *style* parameter determines what type of dot to draw. *STYLE_CDOT* draws a city dot. *STYLE_NDOT* is for the first node dot while *STYLE_NDOT2* is for subsequent node dots.

NetworkElastic

aip.neural.NetworkElastic The *NetworkElastic* class performs all of the work in the EN. Inside of *NetworkElastic* are two private classes, *Node* and *City,* that are used to carry node and city information.

At the top of the *NetworkElastic* file are a number of parameters that control its operation.

A summary of *NetworkElastic* is given in Table 9-5.

addCity(double x, double y) Once the network has been created you need to add cities to it. Each network is only usable for a single city layout. As noted above, the net assumes that the city layout is scaled to a box in the range of –1.0 to +1.0.

reset( ) The *reset* method does not affect the cities. It instead resets the node path and the training parameters.

step( ) At each step, the elastic network matches up nodes and cities with *do_match_nodes()*. It then calculates the attraction between nodes and cities with *do_attract_nodes()* and *do_attract_cities()*.

The network moves the nodes using these attractions and the surface tension using *do_shift_nodes()*.

Once the nodes have been shifted, unused nodes are killed and new ones may be added in *do_grow_die()*.

Once the path is visiting each city, the network drops out of training mode and performs greedy swap exchanges with *do_swap()* until the path is as good as it will get.

Every few passes the nodes are also given a bit of a random kick, with *do_kick()*, to keep the system from falling into a rut.

pathLength( )
generation( ) These methods return the current path length and generation number.

cityIterator( )
nodeIterator( ) These iterator methods give you access to the cities and nodes. These are based on *Point2D.Double*, so they are easy to view and, if needed, manipulate using this standard class.

TSP.java

aip.app.tsp.TSP The behavior of the TSP application should be familiar to you by now. It creates the user interface and then continuously runs the simulation while displaying the results.

Internally, we have added two new and interesting behaviors. We already introduced the parser. The other detail is our use of a file browser dialog to locate the city file.

Once a city file has been read into the parser, it is scaled and sent to the network.

When the *Start* button is pressed, the network resets and the simulation runs until a solution is found.

Additional details displayed to the UI include the generation count and the current path length. Note that the path length starts small and grows as the nodes are stretched out to their cities. After training, there is a brief flash of swapping reduces the path length bit, and then the network stops processing.

Document Processing

All of the networks discussed in the last two chapters operate on vectors in state space, which is fine for colors, inputs from robotic sensors, and positions on a map. But what if you want to process text?

Words

For text and document processing, you need to find a way to map the words or documents into reasonably-sized vectors in state space.

An example of mapping a word onto an SOM-friendly vector is found in Miikkulainen (1993). In his system, he arbitrarily chooses a state-space with seven dimensions. The value of each ordinate in the state vector is based on the coding of the letter at that position in the word. The first ordinate matches the first letter and so forth, for up to seven letters. For words that are less than seven letters long, the remaining ordinates are given the value of zero. For words that are longer than seven letters, the extra letters are truncated, or an abbreviated seven-letter form of the word is used.

The coding value of a letter is based on the darkness of that letter as rendered by, in his case, the Geneva font on the Macintosh. The total number of pixels is divided by the number of black pixels, giving a gray-scale value between zero and one. For example, see the word vectors in Figure 9-20.

Documents

Documents provide a more complex coding problem. The WEBSOM project at websom.hut.fi/websom/ has addressed this task with some success, so we look at their solution (Kohonen, 2000).

A document is composed of words, often hundreds or thousands of words. One way to characterize the document, then, is by the statistical

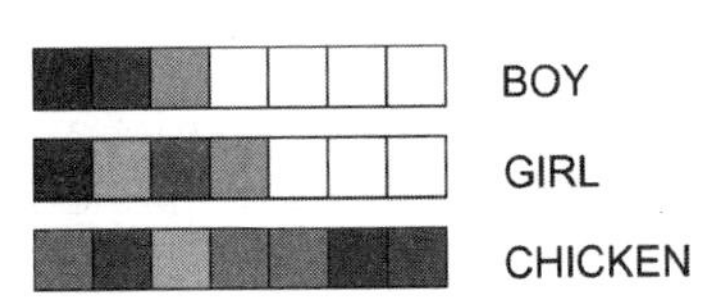

Figure 9-20 Word vector coding

distribution of its words. This distribution takes the form of a normalized histogram. A histogram is a vector with one ordinate slot for every possible word used by the documents. Each time a word is used in a particular document, its slot is incremented by one. This giant vector is then normalized for later processing.

A collection of documents may use thousands of unique words. Some of these words are so common as to provide no special meaning, so they are removed. Other words are so rare as to be of little use differentiating documents, so these are also removed.

Even filtering the two ends of the bell curve, there may still be thousands of words. Each document histogram is a multi-thousand dimension vector, making it impossible to use gracefully.

The next step is to reduce the dimension of the histogram vector. Of course, SOMs provide a dimension reducing service. Or the words could be clustered into groups of similar meaning.

One unusual method of dimensionality reduction is random mapping (Kaski, 1998). In random mapping, you create a rectangular matrix R filled with random values. The height of this matrix is the same as the histogram's length, while the width is the dimension you wish to project down to, typically about one hundred columns.

Each column is normalized to a length of one, making it a unit vector that should be unique from its neighbors. The columns are, in fact, reasonably orthogonal to each other. Multiplying the histogram vector h against the matrix R generates a new vector x that encodes the similarity of h to each column in R. This is our familiar dot-product similarity test, on a grand scale:

$$x = hR \tag{9-13}$$

The new x vector is much smaller than the original histogram, yet it represents the document with little loss of information. Regular neural network techniques can then be applied to it.

Hierarchy and Time

For complex applications, a single neural network will not be sufficient to solve the problem. In some cases, the input vectors grow so large that it becomes impractical to process them with a single network. In other cases, you may find that one network just does not do a good job of representing the data. Or the information is coming into the system as a sequence of inputs and not all at once, and you want to be able to use this time dimension.

This section explores some alternate structures that you can use for your SOMs and other neural networks.

Multi-Modal SOM

In robotics applications in particular, your artificial intelligence is receiving data from several different sensors and other inputs. There may be a ring of sonar or optical distance sensors, a set of bumpers, sound inputs, and of course any commands or internal goals.

Other types of application may have inputs from different modalities as well.

One way to organize the various inputs is to concatenate the input vectors into one monster vector, and send this off to the neural network.

A different, more efficient, method is to send each input modality to its own network, as shown in the simplified SOM in Figure 9-21.

Each input type, *A, B,* and *C,* is presented to its own network. These networks are trained as any regular SOM, but their results are processed by the next SOM layer to create the final output.

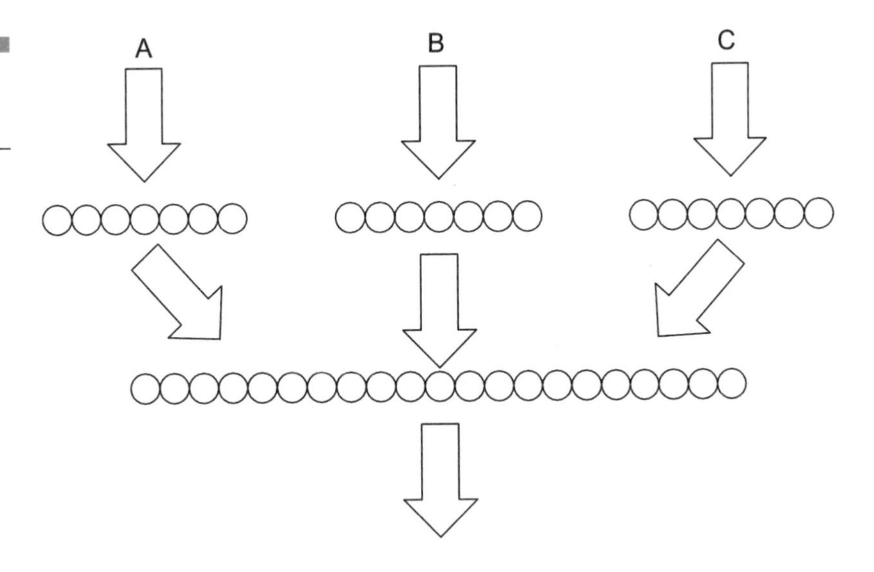

Figure 9-21 Multi-modal network

If this network is a biologically feasible SOM, then there is no one "winner". There will be one or more highly activated neurodes and a number of less active neurodes. The contents of each network are concatenated together making a large vector. If the networks are two dimensional, then each row of each network is also concatenated together. After all, a grid is really just a line that has been folded in on itself.

This giant vector of concatenated SOM results is then passed to another network. This next SOM processes it and calculates a meta-category for the multi-modal data. This multi-model technique is explained in great detail in Briscoe (1997a,b), and is also explored in Wise (2003). A different, yet similar, approach to multi-modal data is found in Ballard (1997).

One problem with this approach is the huge size of the intermediate vectors. These vectors could be reduced by random mapping, as introduced in the document section above.

A different approach is to use the inherent dimension reduction of the SOM when creating the middle-level vector. Using the more traditional winning-node SOM, there will only be one active neurode. With data this sparse, it would be wasteful to concatenate it all together. Instead, we can treat the active neurode as a position in state space, based on its position inside the SOM lattice.

In a line of neurodes, you can consider the left-most neurode to be at position zero, or alternately negative one. The right-most neurode is then at positive one. An active neurode marks a spot somewhere between those two extremes.

If you are using a two-dimensional SOM then you have one scalar value for the X and one for the Y axes of the network. These combine to make a two-ordinate vector.

The vectors of each modality are then concatenated as before and sent to the next level network, as shown in Figure 9-22.

Hierarchical SOM

Another way to reduce the computation involved with a large input vector is to reduce the number of neurodes that must process it. While growing your networks can limit the number of neurodes a minimum, it still creates large, flat networks.

A hierarchical self-organizing map (HSOM) keeps the number of neurodes processing each input to a minimum. Each network in the tree is small, on the order of 2×2 or 3×3 neurodes.

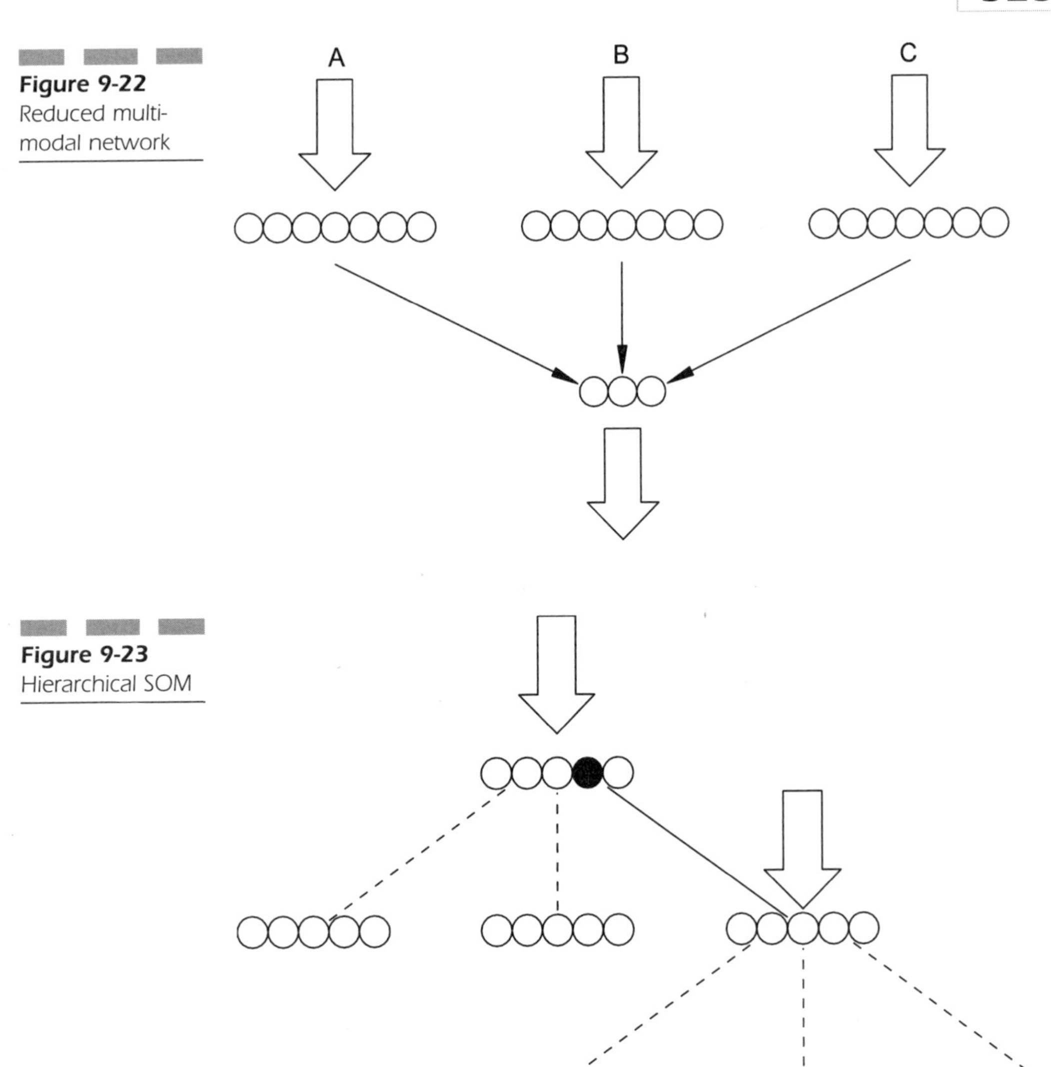

Figure 9-22
Reduced multi-modal network

Figure 9-23
Hierarchical SOM

For each neurode in the top network, there is a sub-network of similarly small size. The input vector is first presented to the top-layer network and a winning neurode, which represents the broad category the input belongs to, is selected. The input is then presented to the network associated with the winning neurode from the previous layer, and so on down the tree until you reach the bottom (Figure 9-23).

The network is shaped like a pyramid. If you piece all of the subnets together, edge to edge, at the bottom you still end up with a large flat network. However, only a small section of it has processed any given input.

The top level network is trained first. Once it stabilizes, you can start sending inputs down to the next levels until they are trained. An example of this type of structure for image compression can be found in Barbalho (2001), and for language processing in Miikkulainen (1993).

A more complex approach adds the grid-growing capability to the hierarchical SOM. This growing HSOM trains one layer until it is stable. Then, when an input arrives that is a particularly poor match in this network, a new sub-network is created under it. Appropriate inputs are then tested against this subnet and it is grown and stabilized. Each neurode in a layer may or may not have a subnet, and each subnet will grow to its most appropriate size.

Of course, to implement this you need a good definition of "stable" and "appropriate". Look to the original paper in Dittenbach (2002) for more details.

Multi-Modal Hierarchical SOM We can be brief here. Given a network as shown in Figure 9-23, the first layer of input can come from one modality. A winning neurode is found and then the input for the second modality is applied to that subnet. An example of this structure in robot control is found in Versino (1996) and Versino (2000). In this example, the first layer processes the sensory input vector and the second layer processes the goal vector.

Time-Series

What is time? And why would we want to incorporate time into our neural network models?

Time provides context. With time, there is a before, a now, and, for creatures with the ability to plan, an after.

The networks explored so far only know about the now. Each input is the only input and it is evaluated against the current state of the system. The learning networks only incorporate time and context to the extent that input events change the state of the system. When the next input arrives, it is compared to a subtly different system than before. Time leaves its mark on any learning system.

However, providing an explicit sense of time gives the network context across a much shorter time scale. An event can be evaluated with reference to the previous event, and even against the effects of events that came before that one.

One way to manage time is to concatenate the input vector from this time slice t to some historical input vectors at, for example, times $t - 1$ and $t - 2$ (Figure 9-24).

At the next input cycle, everything shifts left and the process repeats.

A more efficient method is to move from feed forward networks to recursive networks. Two forms of this were shown in Figures 8-25 and 8-26. This recursive architecture can also work for SOMs, as shown in Figure 9-25.

The input vector is concatenated with a context vector, creating a larger input for the network. This is then processed by the SOM and a winning node is selected. The X and Y position of this node is calculated, creating an output vector. This output is cycled back up to the top to provide context for the next input. In this way, two identical inputs may provide different outputs depending on what came before.

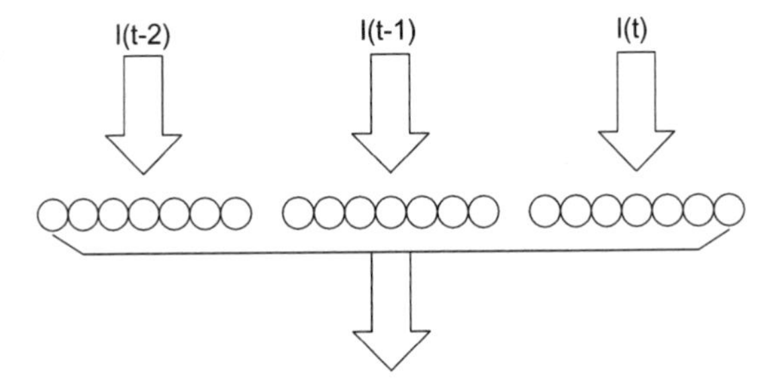

Figure 9-24 Input vectors over time

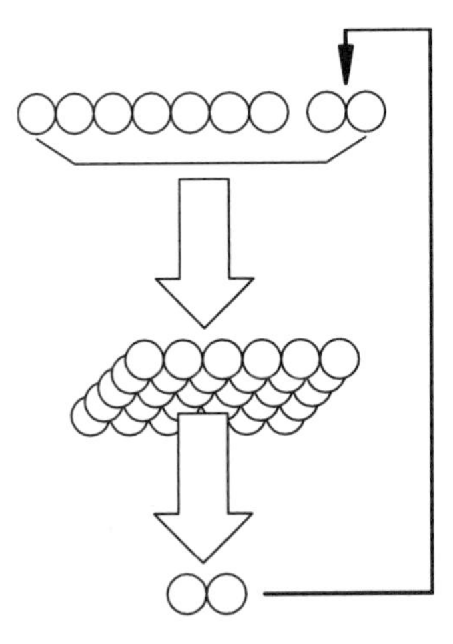

Figure 9-25 Recursive SOM

Figure 9-25 shows a reduced version of the network. Versions using the biologically plausible maps may use the state of all of the neurodes, concatenated into a very large vector for the output and context.

For these methods, though, to preserve your sanity and cycle time, you may want to first reduce the results through random mapping.

For all time series networks, if the output is fed back to the input, like in the Hopfield network, you can train it to reproduce a chain of events.

REFERENCES

Note that most of the research papers referenced in this book can be found via CiteSeer, at www.citeseer.com. This is an amazingly useful research tool.

Abbott, L.F., and Kepler, Thomas B. *Model Neurons: From Hodgkin–Huxley to Hopfield* in Garrido, L. (ed.), *Statistical Mechanics of Neural Networks*, pp. 5–18, Springer, 1990.

Allen, James. *Natural Language Understanding.* Benjamin Cummings/Addison Wesley Publishing Company, 1995.

Altrock, Constantin von. *Fuzzy Logic & NeuroFuzzy Applications Explained.* Prentice Hall, 1995.

Arkin, Ronald C. *Behavior-Based Robotics*. MIT Press, 1998.

Ballard, Dana H. *An Introduction to Natural Computation*. MIT Press, 1997.

Barbalho, José M., et al. *Hierarchical SOM Applied to Image Compression.* Proceedings of the International Joint Conference on Neural Networks, Washington, DC, pp. 442–447, July 2001.

Barr, Avron, and Feigenbaum, Edward A. *The Handbook of Artificial Intelligence*, Volume 1. William Kaufmann, 1981.

Bigus, Joseph P, and Bigus, Jennifer. *Constructing Intelligent Agents Using Java*, second edition. John Wiley & Sons, 2001.

Bolc, Leonard, and Cytowski, Jerzy. *Search Methods for Artificial Intelligence.* Academic Press Ltd, 1992.

Braitenberg, Valentino. *Vehicles: Experiments in Synthetic Psychology*. MIT Press, 1984 (fifth printing, 1996).

Briscoe, Garry. *Adaptive Behavioral Cognition.* Ph.D. Thesis, Curtin University of Technology, Australia, 1997.

Briscoe, Garry, and Caelli, Terry. *ABC: Biologically Motivated Image Understanding* in Caellie, Terry, and Bischof, Walter F., (ed.) *Machine Learning and Image Interpretation*. Plenum Press, 1997.

Brooks, Rodney A. *A Robust Layered Control System for a Mobile Robot*. MIT A.I. Memo 864, 1985.

Buckland, Mat. *AI Techniques for Game Programming*. Premier Press, 2002.

Carver, Norman, and Lesser, Victor. *The Evolution of Blackboard Control Architectures*. Technical Report 92-71, Department of Computer Science, University of Massachusetts, 1992.

Cohen, Paul R., and Feigenbaum, Edward A. *The Handbook of Artificial Intelligence*, Volume 3. William Kaufmann, 1982.

Corkill, Daniel D. *Blackboard Systems* in *AI Expert* 6(9):40–47, September 1991.

Cox, Earl. *The Fuzzy Systems Handbook*. AP Professional, 1994.

DARPA. *Specification of the KQML Agent-Communication Language—plus example agent policies and architectures*. DARPA Knowledge Sharing Initiative External Interfaces Working Group, 1993.

Dautenhahn, Kerstin (1999). *Embodiment and Interaction in Socially Intelligent Life-Like Agents* in Nehaniv , C.L. (ed.) *Computation for Metaphors, Analogy and Agent*, Springer Lecture Notes in Artificial Intelligence, Volume 1562, Springer, pp. 102–142.

DeLoura, Mark. *Game Programing Gems*. Charles River Media, 2000.

Dittenbach, Michael, Merkl, Dieter, and Rauber, Andreas. *Organizing and Exploring High-Dimensional Data with the Growing Hierarchical Self-Organizing Map.* Proceedings of the First International Conference on Fuzzy Systems and Knowledge Discovery, November 18–22, 2002, Singapore.

Dowling, John E. *Neurons and Networks: An Introduction to Neuroscience.* The Belknap Press of Harvard University Press, 1992.

Durbin, Richard, and Willshaw, David. *An analogue approach to the traveling salesman problem using an elastic net method.* Issue 326, Nature, pp. 689–691, 1987.

Dudek, Gregory, and Jenkin, Michael. *Computational Principles of Mobile Robotics.* Cambridge University Press, 2000.

Farley, Jim. *Java Distributed Computing*. O'Reilly & Associates, 1998.

Foundation for Intelligent Physical Agents. *FIPA Abstract Architecture Specification*. Document SC00001L, FIPA, 2002. http://www.fipa.org

Fritzke, Bernd. *Growing Cell Structures—A Self-organizing Network for Unsupervised and Supervised Learning.* TR-93-026, International Computer Science Institute, UC-Berkely, 1993.

Fritzke, Bernd. *Growing Grid—a self organizing network with constant neighborhood range and adaptation strength*. Neural Processing Letters, Vol. 2, No. 5, pp. 9–13, 1995.

Fullmer, David, and Miikkulainen, Risto. *User Marker-Based Genetic Encoding of Neural Networks to Evolve Finite-State Behavior.* Department of Computer Science, University of Texas at Austin, 1991.

Genesereth, Michael R., and Fikes, Richard E. *Knowledge Interchange Format Version 3.0 Reference Manual*. Stanford University Logic Group, Logic-92-1, 1992.

Hodgkin, A.L., and Huxley, A.F. *A Quantitative Description of Membrane Current and its Application to Conduction and Excitation in Nerves* in *Journal of Physiology* 117:500–544, 1952.

Holland, John H. *Adaptation in Natural and Artificial Systems: An Introductory Analysis with Applications to Biology, Control and Artificial Intelligence.* University of Michigan Press, 1975.

International Electrotechnical Commision (IEC). *IEC 1131—Programmable Controllers part 7, Fuzzy Control Programming*. Committee Draft 1.0, 1997. www.fuzzytech.com/e/iec.html

Kaski, Samuel. *Dimensionality Reduction by Random Mapping: Fast Similarity Computation for Clustering.* Proceedings of the International Joint Conference on Neural Networks, 1998.

Katz, Randy H. *Contemporary Logic Design*. Benjamin Cummings/Addison Wesley Publishing Company, 1993.

Kohonen, Tuevo, Kaski, Samuel, et al. *Self organization of a massive document collection*. IEEE Transactions on Neural Networks, Vol. 11, No. 3, May 2000.

Kohonen, Teuvo. *Self-Organizing Maps*, third edition. Springer, 2001.

Kosko, Bart. *Neural Networks and Fuzzy Systems: A Dynamical Systems Approach to Machine Intelligence.* Prentice Hall, 1992.

Kosko, Bart. *Fuzzy Engineering.* Prentice-Hall, 1997.

Kremer, Dr. Stefan C., and Stacey, Dr. Deborah A. *Artificial Neural Networks: From McCulloch Pitts Neurons to Back-propagation*. Notes for class 27-642, University of Guelph, Canada, January 1999. http://hebb.cis.uoguelph.ca/~skremer/Teaching/27642/BP/

Labrou, Yannis, and Finin, Tim. *A Proposal for a new KQML Specification.* Computer Science and Electrical Engineering Department, University of Maryland, TR CS-97-03, February 1997.

Labrou, Yannis, Finin, Tim, and Peng, Yun. *The Current Landscape of Agent Communication Languages*. IEEE Intelligent Systems, Vol. 14, No. 2, March/April, 1999.

Levine, John R., Mason, Tony, and Brown, Doug. *lex and yacc.* O'Reilly & Associates, Inc., 1995.

Li, Hongsing, Chen, C.L. Philip, and Huang, Han-Pang. *Fuzzy Neural Intelligent Systems: Mathematical Foundation and the Applications in Engineering*. CRC Press, 2001.

Linda User's Guide and Reference Manual. Scientific Computing Associates, 2000. http://www.lindaspaces.com

van der Linden, Peter. *Just Java: fourth edition*. Sun Microsystems Press, A Prentice-Hall Title, 1999.

Maass, Wolfgang, and Bishop, Christopher M. (ed.), *Pulsed Neural Networks*, MIT Press, 1999.

Masters, Timothy. *Practical Neural Network Recipes in C++*. Academic Press, Inc, 1993.

Martin, David L, Cheyer, Adam J., and Moran, Douglas B. *The Open Agent Architecture: A Framework for Building Distributed Software Systems* in Applied Artificial Intellignece, Vol. 13 No. 1–2, pp. 91–128, January–March 1999. http://www.ai.sri.com

Michalski, Ryszard S., Bratko, Ivan, and Kubat, Miroslav. *Machine Learning and Data Mining: Methods and Applications*. John Wiley & Sons, 1998.

Miikkulainen, Risto. *Subsymbolic Natural Language Processing: An Integrated Model of Scripts, Lexicon, and Memory.* MIT Press, 1993.

Moriarty, David, and Miikkulainen, Risto. *Evolving Complex Othello Strategies Using Marker-Based Genetic Encoding of Neural Networks*. Technical Report AI93-206, University of Texas at Austin, 1993.

Nilsson, Nils J. *Artificial Intelligence, A New Synthesis*. Morgan-Kaufmann Publishers, Inc., 1998.

Obermayer, Klaus, and Sejnowski, Terrence J. (ed.). *Self-Organizing Map Formation: Foundations of Neural Computation*. MIT Press, 2001.

Oja, Erkki, and Kaski, Samuel. *Kohonen Maps*. Elsevier, 1999.

Pal, Sankar K. and Mitra, Sushmita. *Neuro-Fuzzy Pattern Recognition: Methods in Soft Computing*. John Wiley & Sons, Inc, 1999.

Prusinkiewicz, Przemyslaw, and Lindenmayer, Aristid. *The Algorithmic Beauty of Plants.* Springer, 1991.

Rabin, Steve. *Designing a General Robust AI Engine* in DeLoura, Mark (ed.) *Game Programming Gems*, Charles River Media, Inc, 2000, pp. 221–236.

Rabin, Steve. *AI Game Programming Wisdom*. Charles River Media, 2002.

Russell, Stuart, and Norvig, Peter. *Artificial Intelligence: A Modern Approach*. Prentice Hall, 1995.

Schaffer, J. David, and Morishima, Amy. *An Adaptive Crossover Distribution Mechanism for Genetic Algorithms* in Grefenstette, John J. (ed.), *Genetic Algorithms and Their Applications: Proceedings of the Second International Conference on Genetic Algorithms*, Lawrence Erlbaum Associates, 1987.

Schraudolph, Nicol N. and Belew, Richard K. *Dynamic Parameter Encoding for Genetic Algorithms*, UCSD Technical Report CS 90-175, University of California, San Diego, July 20, 1992.

Spatz, Chris. *Basic Statistics: Tales of Distributions*, 6th edition. Brooks/Cole Publishing Company, 1997.

Sutton, Richard S. and Barto, Andrew G. *Reinforcement Learning, and Introduction*. MIT Press, 1998. http://envy.cs.umass.edu/~rich/book/the-book.html

Tozour, Paul. *Introduction to Bayesian Networks and Reasoning Under Uncertainty* in Rabin, Steve. *AI Game Programming Wisdom*. Charles River Media, 2002.

Tracy, Kim W., and Bouthoorn, Peter. *Object-Oriented Artificial Intelligence Using C++*. W.H. Freeman and Company, 1996.

Versino, Cristina, and Gambardella, Luca Maria. *Learning Fine Motion in Robotics: Experiments with the Hierarchical Extended Kohonen Map*. Proceedings of the International Conference on Neural Information Processing, Hong Kong, September 24–27 , Vol. 2, pp. 921–925, 1996.

Versino, Cristina, and Gambardella, Luca Maria. *Learning Fine Motion in Robotics: Design and Experiments.* Recent Advances in Artificial Neural Networks, Design and Application, CRC press, pp. 127–153, 2000.

Weiss, Gerhard (ed.). *Multiagent Systems: A Modern Approach to Distributed Artificial Intelligence.* MIT Press, 2000.

Wise, Edwin. *Applied Robotics II.* Delmar Learning, 2003.

Witten, Ian H., and Frank, Eibe. *Data Mining: Practical Machine Learning Tools and Techniques with Java Implementations.* Morgan Kaufmann Publishers, 2000.

INDEX